Managing Electrical Safety

James H. Wiggins, Jr., CSP

ABS Consulting
Government Institutes
Rockville, Maryland

ABS Consulting

4 Research Place, Rockville, Maryland 20850, USA
Phone: (301) 921-2300
Fax: (301) 921-0373
Email: giinfo@govinst.com
Internet: http://www.govinst.com

Printed in the United States of America

Library of Congress Cataloging-in-Publication Data

Wiggins, James H.
   Managing electrical safety / James H. Wiggins, Jr.
      p. cm.
   Includes index.
   ISBN 0-86587-889-7
   1. Electric engineering--Safety measures. I. Title.

TK152.W54 2001
621.319'24'0289--dc21   2001040835

Contents

Electrical safety should be of great concern to safety professionals. Electricity is prevalent throughout our society, but industrial machinery often utilizes voltages much higher than the household or office circuits with which most people are familiar. However, unlicensed electricians often perform electrical maintenance in an industrial setting. Employees are often required to work around exposed live parts such as crane rails, open electrical control cabinets, or near overhead lines or service drops. The potential for electric shock and arc flash burns can be very high. From 1999 to 2000, the Occupational Safety and Health Administration (OSHA) cited 7,847 electrical safety violations. Only 178 of those citations were for utility-type work. The breakdown of electrical safety violations for the period was as follows:

Standard Violated	No. of Citations
• Wiring Methods, Components, and Equipment for General Use (29 CFR 1910.305)	3232
• General Requirements (29 CFR 1910.303)	2478
• Wiring Design and Protection (29 CFR 1910.304)	953
• Selection and Use of Work Practices (29 CFR 1910.333)	358
• Use of Equipment (29 CFR 1910.334)	269
• Electric Power Generation, Transmission, and Distribution (29 CFR 910.269)	178
• Training (29 CFR 1910.332)	155
• Hazardous (Classified) Locations (29 CFR 1910.307)	139
• Safeguards for Personnel Protection (29 CFR 1910.335)	78
• Specific Purpose Equipment and Installations (29 CFR 1910.306)	6
• Special Systems (29 CFR 1910.308)	1

This book will assist safety professionals manage their electrical safety programs. The focus will be on typical industrial environments, not the electrical utility industry. Some of the electrical activities to be managed in an industrial facility include operational troubleshooting, installation of equipment, and maintenance of equipment and facilities. The topics outlined in the table of contents will only be discussed to the degree needed to assist the safety professional to formulate the right questions to ask of their engineering support personnel, operational management, and facilities management. The goal of this book is to assist the safety professional to ensure that adequate electrical safety will be managed continuously.

First, the book discusses a model safety management process. The process consists of a structure and a procedure to manage the risk of injury from electrical hazards. The concepts of risk management and acceptable level of risk are developed in general and specifically as they apply to electrical safety. Second, the book outlines the development of an electrical safety program including hazard identification and control. As a further discussion of hazard identification, the conduct of job hazard analysis (JHA) is discussed in general, as well as specific issues that arise when incorporating electrical maintenance and other non-routine tasks into the JHA. Because a personal protective equipment (PPE) assessment is often performed in conjunction with a JHA, PPE for electrical shock and arc flash is discussed in a fair amount of detail. Next, the book turns to a discussion of procedures for ensuring equipment is de-energized before work (lockout/tagout) and for safely planning work on or around exposed live parts.

Part of the model safety management process is periodic review of program performance, feedback, and continuous improvement. Therefore, the book concludes with a discussion of reviewing proposed changes to facilities and equipment to ensure that, as systems are upgraded, risk remains below the acceptable level. Considerations reviewed here include adequate clearance, electrical installation methods for hazardous areas, grounding, specifications for safe systems, ease of lockout/tagout, and location of disconnects.

Finally, periodic electrical safety inspection will ensure that the programs are working efficiently and effectively. And preventive maintenance is discussed to ensure that degradation of electrical systems does not compromise their safety. At the end of the book, an appendix gives a quick overview of the basics of electricity and its hazards—including electric shock, arc flash, ignition of flammable and combustible material, static and low frequency field exposure, and non-ionizing radiation exposure.

James H. Wiggins, Jr. CSP

Mr. Wiggins is a certified safety professional with a specialty in System Safety Aspects. He graduated from the University of Texas at Austin with a Bachelor of Science in physics and went on to become a nuclear power trained, submarine-qualified officer in the US Navy. He has worked on the Space Shuttle and Space Station programs, performing reliability and hazard analyses and managing the Software Safety Group on contract at NASA's Johnson Space Center.

Currently, Mr. Wiggins is a Senior Safety Consultant for Clayton Group Services, Inc. in the Detroit, Michigan area. Mr. Wiggins provides expertise to his clients in the areas of Electrical Safety, Fire Prevention, Process Safety Management, Machine Safeguarding, Control of Hazardous Energy, Confined Space Entry, and Ionizing Radiation Safety. He has presented papers at two international safety, reliability, and quality conferences and has published several articles in monthly periodicals.

Chapter One

Safety Management

Overview

To maintain risk at an acceptable level—this is the goal of safety management. To achieve the goal, one must have a proactive safety process—an administrative system for analysis, control, and feedback of risk. It is a process of continuous improvement. No single safety program or even the sum of all safety programs can accomplish the goal. The safety management process provides a structure whereby the appropriate levels of management have adequate information to make good business decisions that reduce risk to, and maintain risk at, an acceptable level.

Management structure alone is insufficient to accomplish the goal. Every company and facility has a management structure. The management structure can facilitate or hinder the pursuit of the goal. What are the appropriate levels of management responsible to make specific decisions? Do the responsible persons have the authority to make those decisions? Do these levels of authority have the information, training, and tools they need to make those decisions? Are they held accountable for the decisions they make? If the risk is unacceptable, the reasons usually are rooted in a negative answer to one of the questions above.

However, even the best management structure cannot perform a function that it was not set up to perform. Managers, from the executive to the cell leader, must clearly delegate safety responsibilities and authority to their subordinates and hold the subordinates accountable for the performance of those safety responsibilities. Additionally, the management structure must implement a process for routinely obtaining and communicating adequate information about the current or proposed level of risk.

Modeling the Safety Management Process

In developing a proactive safety process, the goal is to integrate safety into the management structure, safety programs, and production processes. The safety process is similar to a quality process. If your facility has a quality process, the safety process should closely parallel it.

The process by which safety is managed can be modeled as a continuous process of identifying safe or unsafe conditions, selecting controls that reduce the risk to an acceptable level, and modifying procedures to prevent the occurrence of similar conditions without adequate controls. Safety indicators, objectives, or goals must be measurable and well-defined. Just as quality is an integral part of production, so too is safety. The safety objectives must be proactive. Do not rely on Lost Work Day Incidence Rates (LWDI) or numbers of accidents to indicate the degree to which the safety process is successful. The bottom line is with how much risk is the facility operating and whether or not that level of risk is acceptable to management. Some suggestions for safety indicators are:

- Incidence of personnel wearing hearing protection or eye protection in areas where it is required

- Results of quantitative safety analyses of systems, machines, or production lines

- Incidence of poor housekeeping around electrical equipment, open flames, hot work, aisles, and walking/working surfaces

- Incidence of slippery floors, tripping hazards, and inadequately guarded floor and wall openings

- Incidence of personnel using proper electrical PPE

- Incidence of personnel using machine-specific lockout procedures

Measurement can take the form of periodically scheduled inspections of work areas or equipment, observations of performance of specific tasks, and procedure and program reviews. Measurement can also take the form of qualitative or quantitative determination of residual risk associated with all proposed and new or modified equipment. Measurement can also take the form of monitoring for adequacy of engineering controls. An example would be monitoring for flammable gases or vapors to determine the effectiveness of ventilation in preventing flammable atmospheres in the vicinity of ordinary electrical installations.

Analysis of measurements is performed to determine trends and areas needing improvement. The results of this measurement analysis must be communicated to all levels of management and affected personnel. But the process must not stop with communication. Steps must be taken to correct any deficiencies. The appropriate level of management must determine what actions will be taken, assign responsibility for completion of the actions, and track the actions to completion. Tracking action items to completion is essential, as it is the means by which upper levels of management can hold their subordinates accountable for their responsibilities.

Management of Risk

The safety management process endeavors to identify risks and control hazards to reduce risk to an acceptable level. This lofty goal is often thwarted by the realities of human behavior. People are independent creatures who exercise their free will and often choose to disobey safety rules or fail to use common sense. The focus of the safety professional too often becomes a response to the realization of risk in the workplace. Accidents happen; accidents must be investigated; workers' compensation claims must be managed; limited duty and return to work programs must be supervised; employees must be retrained in safety rules, practices, and procedures. Where in the midst of this flurry of activity is the management of risk? A process is needed by which the hazards are identified.

The hazards must be identified for each machine or piece of equipment and for each task. Many safety programs address these issues: machine safeguarding, electrical safety, hearing conservation, chemical exposure control, hazard communication, respiratory protection, personal protective equipment, blood-borne pathogen exposure control, control of hazardous energy, permit-required confined space entry. However, in the face of all these programs that must be administered, where is the responsibility to manage risk? Where is the determination of what an acceptable level of risk is?

In some cases, the acceptability of risk is dictated by government regulation. But in many cases, the government regulation is a performance-based standard. The unacceptability is only proven after an accident is realized. This is not a proactive approach. A company can better define an acceptable level of risk on the basis of sound business criteria. How much money can the company afford to waste on managing accident cases, including legal fees? Does an existing level of risk create ill will between the company and a contracted labor union? How much bad publicity can the company afford because of a fatal accident, an explosion, or an environmental impact? These questions are usually not within the ability of the safety professional to answer. The highest levels of management must make the determination of what is an acceptable level of risk. Once determined, everyone in the company must work to reduce the risk and keep it at or below the acceptable level in accordance with the defined risk management process.

Risk Management Process

The risk management process is the way in which a company will identify the risk it currently operates under and the means by which it will decide how to reduce that risk to an acceptable level. The risk management process will be proactive if it prevents accidents and injuries, manages change to maintain risk below the acceptable level, and provides for continuous improvement. However, for existing facilities and plants, the risk management process must also identify existing safety problems in sufficient detail to correct them and reduce the existing level of risk to an acceptable level. The risk management process for any given facility will be unique to the management structure, breadth and depth of knowledge within the management team, and complexity and size of operations. However, all risk management processes should have the following major elements:

- Hazard identification
- Hazard control
- Determination of acceptability of residual risk
- Program evaluation and improvement

Hazard Identification

Identifying hazards is not difficult if one knows what to look for (see Figure 1.1). There are regulations, codes, and industry consensus standards that address most safety topics. For electrical safety, the *National Electrical Code®* (NEC®) and *National Electrical Safety Code®* (NESC®) are often adopted as state and local codes. Federal or state occupational safety and health regulations on electrical safety and lockout/tagout are pertinent to electrical installations and maintenance. The National Fire Protection Association (NFPA) standards on *Electrical Safety Requirements for Employee Workplaces* (NFPA 70E) and the *Flammable and Combustible Liquids Code®* (NFPA 30) are two examples of industry consensus standards that contain electrical safety requirements. Electrical safety requirements for control panels and other parts of equipment are governed by the *Electrical Standard for Industrial Machinery* (NFPA 79).

- Identify applicable regulations, codes, and standards.
- Develop a checklist.
- Conduct a detailed safety inspection of equipment (e.g., electrical installations).
- Conduct an audit of safety rules, procedures, and practices.

Figure 1.1: Hazard Identification

The federal Occupational Safety and Health Administration (OSHA) and the American National Standards Institute (ANSI) provide safety requirements for various machines, including recommended practices for machine safeguarding, controls, electrical requirements, and ergonomic design. The ergonomic design of equipment is necessary to protect operators and maintenance personnel who work on the equipment. The *Life Safety Code®* (NFPA 101) or the local building code contains requirements for egress, minimum width of aisles, minimum headroom, and acceptable height of guardrails. These are but a few of the sources for safety requirements.

- Mechanical motion
- Gravitational potential
- Spring potential
- Electricity, static charge, and electric potential
- Magnetic fields
- Fluid flow, pressure (hydraulic energy, pneumatic energy, steam)
- Chemicals
- Temperature extremes
- Ionizing and non-ionizing radiation

Figure 1.2: Types of Energy

To identify hazards, define the scope of the project carefully. Do not try to inspect too much at once. It is difficult to keep all the regulations in mind while performing an inspection. Use the applicable requirements to develop a checklist of what to look for. Then, conduct an inspection of equipment and facilities using that checklist. See Chapter 7 for a sample checklist for electrical installations.

Another way to identify hazards is to look for energy of all types (see Figure 1.2). This method is particularly useful when evaluating machines to develop lockout/tagout procedures. By looking at energy sources, stored energies, and ways that those energies can be released unexpectedly, most electrical hazards will be readily identified.

One can also evaluate a system using the concept of generic hazards. A list of generic hazards is an attempt to group hazards into hazard categories. Then, one needs to determine if the hazard is potentially present in the system or equipment. When identifying hazards, one does not consider controls that are in place. The use of generic hazards for hazard identification has been used in the aerospace and defense industries for decades. There is no one list of generic hazards that is complete and universally applicable. The list usually evolves with experience and is frequently colored by the preferences, likes, and dislikes of the person developing the list. One extremely general list of generic hazards is given in Figure 1.3. The idea behind generic hazard categories is to get the inspector to think from a variety of perspectives. This greatly helps to minimize the chances of omitting a significant hazard.

- Mechanical motion
- Slips, trips, and falls
- Colliding with protrusions
- Falling objects
- Release of fluid under pressure (includes compressed air and steam)
- Impingement by projectiles or entrained particles
- Electrical shock, arc flash
- Fire, explosion
- Oxygen deficiency
- Chemical exposure
- Temperature extremes
- Ioninzing and non-ionizing radiation

Figure 1.3: Generic Hazard Categories

Once the hazards are identified, they should be documented. Since the idea is to determine overall risk, the hazard's severity and likelihood should also be determined. Risk is a combination of severity and likelihood. A specific hazard usually has many possible degrees of severity. For example, an automobile accident can be nothing more than a fender-bender or it can be deadly. For each degree of severity, there is a separate likelihood. For example, it is much more likely for an automobile accident to result in minor damage and injury than it is to result in a fatality. The overall risk posed by a hazard is therefore the sum of the varying degrees of severity times their likelihood. However, unless a quantitative risk analysis is needed, identifying the worst-case combination of severity and likelihood) is sufficient.

A subjective, qualitative determination of risk is accomplished by defining the degrees of severity and likelihood to be used and applying those definitions to the hazards identified (see Figure 1.4). When defining the degrees of severity and likelihood, consider the levels of risk that have been deemed acceptable by management. Make sure that the definitions of severity and likelihood will allow discrimination between hazards with acceptable and unacceptable risk. The definitions of the degrees of severity and likelihood are a starting point and can be modified to fit the corporate culture and goals.

The risk posed by the hazard should be determined without regard to controls that might minimize severity or likelihood. Then, any controls already in place should be identified. Hazard controls are not 100% reliable. Hazard controls reduce risk. They reduce the probability of an accident, or they mitigate the severity of an accident. Residual risk is the risk that remains with hazard controls in place. This is the risk that the facility currently operates under. Document the risk of the hazard with and without current controls.

Severity

I   Catastrophic—accident could cause potential loss of life, permanent disability, and/or destruction of equipment. (Forces capable of destroying equipment could potentially cause loss of life, depending on exposure.)

II   Major—accident could result in temporary disability (i.e., less than one year).

III Minor—accident could require medical treatment with no expected disability

IV Negligible—accident would not result in reportable injuries

---

Likelihood

A   Very likely—expected monthly.

B   Likely—expected once per year.

C   Unlikely—expected once every five years.

D   Very unlikely—expected less than once every five years.

Figure 1.4: Degress of Severity and Likelihood

Hazard Control

For each hazard identified with an unacceptable level of residual risk, a better control is needed. The recommended control must reduce the risk to an acceptable level. The types of controls recommended will follow the hazard reduction precedence sequence (HRPS), which is presented in Chapter 3. Hazards with the greatest risk (or highest priority) will receive the most reliable control means practically achievable.

For each hazard identified, document any recommendations for reducing risk to acceptable levels. It is a good idea to determine what the residual risk would be after implementing the recommendations. This number will help management determine the worthiness of the recommendation in reducing risk.

Determination of Acceptability of Residual Risk

These two factors (severity and likelihood) are combined to determine priority based on overall risk. The priority hazard control matrix is shown in Figure 1.5 with the highest priorities having the smallest numbers. Modifying the priority matrix is permitted. An alternative to the priority matrix is a risk matrix where the higher the number, the greater the risk. The numbers are useful only in so far as they provide information to decision-makers to allow them to manage risk.

Figure 1.5: Priority Matrix	A. Very Likely	B. Likely	C. Unlikely	D. Very Unlikely
I. Catastrophic	1	2	4	6
II. Major	3	5	7	9
III. Minor	8	10	11	12
IV. Negligible	13	14	15	16

In order to manage risk to an acceptable level, management must compare the current level of risk to what has been defined as acceptable. Where the risk is high, controls need to be found that would lower that risk. However, some controls that could eliminate the hazard or reduce it to an acceptable level may be economically infeasible. Therefore, some creativity is needed to find solutions that reduce the risk as much as possible while keeping costs down. To make such decisions objectively, assigning a dollar value to risk is sometimes helpful. A dollar value can be assigned to various hazard results, even death. But one should remember that there are many hidden costs to accidents. See the discussion later in this chapter on Safety's Contribution to the Bottom Line.

With a dollar value representing severity of a hazard and a likelihood expressed as an annual expectation, one can calculate the average annual expected cost of accidents. This quantitative approach to risk analysis allows a cost benefit analysis. If a specific hazard control recommendation is implemented, it will reduce the risk by a certain amount. A dollar value can be assigned to the risk before and after implementing the control. The difference is the benefit of implementing the recommendation expressed in dollars. This savings can be weighed against the cost of implementing the recommendation. When taken over the life expectancy of new or modified equipment, the savings can often be shown to be worth the cost. When administrative controls such as procedures and training are recommended, the cost is lower, but the benefits usually are too. Therefore, a cost/benefit analysis is usually very helpful in making management decisions.

Program Evaluation and Improvement

Once the program is in place, the safety professional should evaluate its effectiveness on a regular basis. An effective program will sell itself and maintain the support from upper management needed to implement a safety management program. Additionally, an audit program will identify weaknesses or inefficiency in some areas that need improvement. Personnel may need additional training or education to be able to contribute more to the overall safety program. Procedures may need to be updated, or new procedures may be necessary. These changes to the program must be incorporated into the audit requirements. Continuous improvement is the goal of periodic evaluation. Therefore, the safety professional should enlist other stakeholders in the process. Weekly safety inspections by supervisors over their own areas, monthly safety committee inspections, and annual audits keep the management team focused on safety. Auditing and inspection is required and/or needed in many areas of safety. Efficiency is found when all the various programs are combined as an integrated safety management program so that there is no duplication of efforts.

Integrating Safety Programs

Many of the safety programs required by OSHA are interrelated. For example, the emergency action plans required by OSHA are interrelated to Hazardous Waste Operations and Emergency Response requirements, Fire Brigade requirements, Medical and First Aid requirements, Process Safety Management (PSM) requirements, Fire Extinguisher requirements, and Confined Space Entry Requirements. Other regulatory agencies such as the US Environmental Protection Agency (USEPA) and the Department of Transportation (DOT) have similar or overlapping requirements with OSHA.

For example, the USEPA also requires emergency response plans for spills or releases of hazardous chemicals, and OSHA and USEPA have overlapping requirements for processes involving highly hazardous chemicals (e.g., PSM and the Accidental Release Provisions of the Clean Air Act or Risk Management Plan (RMP)). The DOT requirements for marking hazardous materials is enforced by OSHA and must be integrated into the overall chemical management procedures at the facility and the Hazard Communication program.

Electrical Safety

Of special interest to electrical safety are the overlapping requirements for control of hazardous energy or lockout/tagout. OSHA's electrical safety requirements for working on de-energized equipment (29 CFR 1910.333), the Control of Hazardous Energy Standard (29 CFR 1910.147), and the Confined Space Entry standard (29 CFR 1910.146) all have requirements for controlling hazardous energy. The similarities and differences between the requirements for electrical maintenance on de-energized equipment and the lockout/tagout standard will be discussed in Chapter 4. Additionally, the requirements for personal protective equipment (PPE) assessment (29 CFR 1910.132), Electrical Protective Devices (29 CFR 1910.137), Selection and Use of [electrical] Work Practices (29 CFR 1910.333), and Safeguards for Personnel Protection [from electrical hazards] (29 CFR 1910.335) are closely intertwined.

Accident/Incident Investigation

Another program that needs to be incorporated into the safety management program is Accident/ Incident Investigation. Procedures for accident investigation are required under the PSM and RMP standards and they are inferred from OSHA's record-keeping requirement for details concerning the accident to be documented. Accident/Incident Investigation should incorporate investigation of near misses, fires, and arc flashes, even when no injuries occurred. The reason to investigate such incidents is to find the elusive "root cause" so that similar accidents can be prevented.

Root cause is defined as the original, most significant, or sufficient cause of the accident. In other words, it is the cause that if eliminated would have prevented the accident. There are many approaches or methods for determining root cause, but they all rely on the subjective opinions of the investigator. It is sufficient for safety management purposes to find a cause that can be eliminated and thus prevent future similar accidents. That is the test of whether the root cause has been found.

Too often the root cause is found to be "human error" or an unsafe act. But since human error and the propensity for people to perform unsafe acts from time to time cannot be eliminated, this is not a very useful root cause. The point of accident and incident investigation is to reduce risk. If this cannot be achieved with the root cause identified, keep looking and digging for some other causal factor that can be controlled.

The accident/incident investigation should verify that the Job Hazard Analysis (JHA) identified the hazard involved in the accident. If it did not, the JHA should be updated. The control method for the hazard as documented in the JHA should also be reviewed for its adequacy. If the control was and is considered adequate, the accident and future accidents like it are considered acceptable to management. If the risk of accident is too great, additional or different controls should be developed for the hazard.

Tying Safety Programs into the Safety Management System

As stated before, safety management is much more than a compilation of safety programs. The concepts discussed above must be incorporated into all safety programs and all types of management decisions. The focus must be on reducing risk to an acceptable level. With the management concepts discussed above in mind, each safety program should be reviewed and improved, incorporating the elements of:

- Hazard identification
- Hazard control
- Determination of acceptability of residual risk
- Program evaluation and improvement

Where a routine safety inspection can uncover hazards in multiple program areas, those program requirements can be merged into a single inspection tool or checklist. The management system that reviews the recommendations for improvement and reduction of risk should be the same regardless of the safety program being addressed. The dynamic program model described above is the model for all successful ongoing programs.

Dynamic Program Model

The dynamic program model is a cyclical or repeating process where continuous improvement is generated and the program goals are accomplished.

Management is a cyclical process whereby authority is delegated, responsibility is undertaken, and accountability is obtained. How well the subordinate performs determines how much authority will be granted in the future. The responsibility is performed with enough supervision to ensure its accomplishment and with enough delegated authority to teach leadership.

Communication is also a cyclical process whereby programs and procedures are developed and communicated through training, and the effectiveness of communication is determined based on implementation of the program elements and procedures. Training is essential to communicate the goals and procedures and to provide the necessary tools to accomplish the goals. Retention of knowledge and safe behaviors are observed to determine to what degree the communication has been successful. How well people remember and follow safe procedures determines the frequency, content, and format of future communication. The program becomes an essential part of the company culture.

Implementation of safe practices is accomplished through a cycle of setting requirements, measuring performance against those requirements, analyzing the measurements to determine where performance is deficient or superlative, and providing feedback to change the requirements. The requirements should provide a challenging yet realistic set of goals. How well the program implementation meets the requirements determines the future direction of the safety program.

Management, communication, and implementation are intertwined, cyclical processes that make up the structure of the safety management system.

Safety Organizations

The organizational structure and safety's position in it reflects management's commitment to safety. By placing safety as a direct report to the senior facility manager, management establishes that safety is his/her ultimate responsibility (see Figure 1.6). It ensures that the safety professional has a direct line of communication to the senior decision-maker.

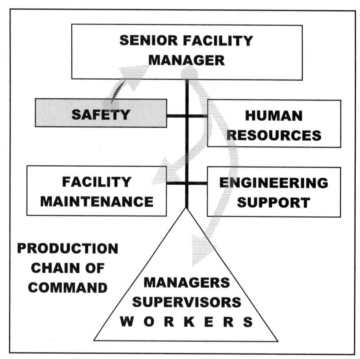

Figure 1.6: Safety as a Direct Report

Often the safety professional is not a direct report to the senior facility manager. Safety is frequently a function of the Human Resources department (see Figure 1.7). As a subordinate of the Human Resources Manager (HRM), the safety professional reports to a person who does not necessarily have any particular knowledge or experience in safety and who is not a decision-maker within the production chain of command. Even when the HRM is the safety professional, the breadth of his/her other duties often interferes with managing safety effectively. The HRM is typically responsible for payroll, benefits, worker's compensation management, labor relations, and sometimes accounting. To add safety to the list of duties may overburden the HRM unless there is sufficient staff to perform much of the work. Why do some organizational structures place safety in the

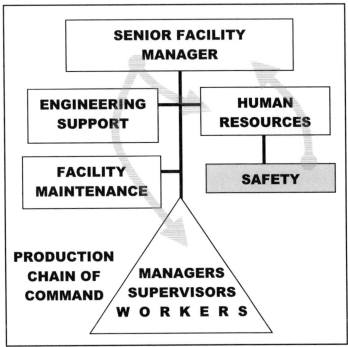

Figure 1.7: Safety as a Function of Human Resources

human resources department? Tradition. Safety at the facility or parent company may have grown out of an employee-relations effort, undertaken and directed by the human resources department. Therefore, safety has always been part of that organization. The problem is that safety and risk management are about how the production department does business. Safety is not primarily an employee-relations program.

Because of the highly technical nature of safety and how it affects the design of facilities and systems, the safety professional is sometimes located in the facilities/maintenance department or engineering support department. These are better choices for the location of the safety support function. However, unless the head of facilities or head of engineering is the safety professional, these organizations still have communications issues. It will be difficult for the safety professional to guide the senior facility manager in his/her decisions on safety and risk management unless he/she is a direct report. It is a conflict of interest for the engineering manager to report to the senior facility manager safety problems engineering was supposed to have previously eliminated or controlled. One way around this issue, regardless of the safety professional's location in the organization, is to make the safety professional a member of the senior facility manager's staff. The safety professional should attend all staff meetings and should be able to speak his/her mind as any other staff member. The senior facility manager must encourage open communication so the right decisions can be made.

Safety management may also break down when there is inadequate management support for safety goals. Without management support, the staff support function of the safety professional is

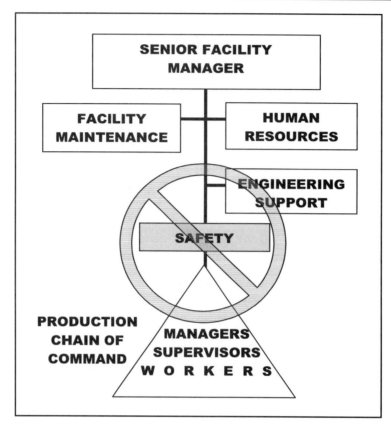

Figure 1.8: Safety as Not a Part of Production Chain of Command

underutilized and viewed with suspicion or as an impedance to productivity. Since the safety professional is usually not in the production chain of command, he/she has no authority to enforce safety procedures and safe work practices (see Figure 1.8). Enforcement is a line supervisor responsibility. If the safety professional were in the production chain of command, it would be a conflict of interest to give production a bad safety report.

The third place safety management breaks down is where upper management fails to delegate safety responsibility throughout the organizational structure. Safety is everyone's job. Even if the safety professional reports directly to the senior facility manager as a member of staff, he/she is not responsible for the safety of the facility. The employees do not report to the safety professional. They report through their supervisors and managers to the senior facility manager. This production chain of command is responsible for the safety of the workplace, and they should be made aware of that responsibility. The senior facility manager should give them the authority and hold them accountable for their safety performance. It is the safety professional's job to facilitate the chain of command in performing their safety responsibilities. The safety professional provides information, guidance, assistance, and coordination of effort. He/she is often the only person in the management structure with any knowledge of safety regulations and standards. Therefore, his expertise and effort are essential for the management team to accomplish their safety goals.

Safety's Positive Contribution to the Bottom Line

Safety professionals are often asked to justify the expense of safety programs and even their own jobs. The problem is that management often sees safety as a staff function that does not contribute to the company's bottom line. This is simply not true. But the safety professional often does not know how to quantify that contribution.

While the safety function may not actually make money for the company, the safety function minimizes waste. By reducing the risk a company operates under and the associated costs, safety professionals and safety programs can be seen to have a positive contribution to the company's bottom line. The key to quantifying the cost of risk is to be thorough in identifying indirect costs.

An accident results in both direct and indirect costs. Direct costs include worker's compensation and property damage. Other costs include lost production capacity while equipment is being repaired and lost productivity when less trained employees have to perform the job of an injured worker.

Direct costs include wage losses, medical expenses, administrative expenses, and employer costs. According to the National Safety Council publication, Injury Facts®, 1999, the average cost of a burn was $8,429 in 1996-1997. The average cost of other trauma, including electrocution, was $15,995 over the same period. The cost of a death in 1998 was $910,000, and the average cost of a disabling injury was $28,000 for the same year. These costs account for only some of the direct costs of the accident. For example, in these statistics, damage to industrial equipment is not included (although the cost of motor-vehicle damage is included).

In an article published in *Professional Safety*, "What Do Accidents Truly Cost?," Jeffrey E. LaBelle, April 2000, 38-42, a list of indirect costs includes the following:

- Time spent away from the job, including travel to/from doctor
- Inefficiency during restricted work
- Supervisor time in consultation with injured worker
- Lost productivity, lost production
- Time spent on incident review
- Time spent by Human Resources department managing case back to 100% duty and record-keeping
- Cost to hire or bring in temporary or permanent replacement
- Training of new or replacement worker, and retraining of injured worker
- Legal costs

Think for a moment about productivity costs. If an employee works overtime to cover an injured worker, the company may pay him time and a half. But studies have shown that overtime results in only meager increases to production. The company has to pay 6 hours wages for 4 hours overtime. At 50% productivity, the 4 hours of overtime are equivalent to only 2 hours at the full production rate. The company also pays the injured worker a portion of his wages according to the company's benefits and worker's compensation laws. Thus, the company could end up paying more than 3 times the original wages to maintain productivity at the pre-accident levels.

If the company instead hires a temporary employee to fill in for the injured worker, the company does not have to pay time and a half, but it does have to pay wages for a lower production rate until the temporary employee learns the job. And the company must still pay the injured worker. Obviously, the situation is not economical. In addition, the temporary employee has not had the safety training of the other employees and is more apt to have or contribute to another accident. It is easy to see that the costs of an accident can mount up quickly.

More insidious costs include increases in medical benefits premiums, typically paid in large part by the company, and claims management costs. The burden of managing return to work programs, worker's compensation claims, and benefits administration is placed on the staff of the human resources department and safety department. This burden requires a number of employees or outsourced contract employees. If there were fewer accidents to manage, costs would be reduced. The amount of cost reduction varies greatly from company to company depending on organizational structure, contractual arrangements, and the accident rate.

Real costs mount up quickly if a fatality occurs. Regardless of worker's compensation laws, many states allow lawsuits to be brought against a company by the families of workers killed or seriously injured on the job through wrongful death or negligence. Even if a company is in the right, the costs of defending against lawsuits can be astronomical.

More costly is bad press that can create a stigma surrounding a company's products and services. Sales or revenue lost by bad press over a serious accident can lead to bankruptcy. It usually takes an accident of catastrophic proportions to create such bad press, but it depends on the size of the community, the impact the accident has on the community, and the level of hype created by the media.

When determining the contribution the safety professional has made to the bottom line, calculate the reduction of risk and the costs associated with that reduced risk. On average, for every accident that is prevented, tens of thousands of dollars are saved.

Chapter Two

Developing an Electrical Safety Program

Scope of the Task

Managing electrical safety is about planning for routine and non-routine tasks that have electrical hazards. Every type of industry and operation has electrical hazards. Plugging and unplugging electrical appliances can be electrically hazardous. Changing a lightbulb or a fluorescent fixture ballast can be hazardous. Painting or cleaning around exposed live parts is particularly hazardous. If an ordinary-rated cord becomes wet, even operating a portable electric tool can result in shock. From the office to the production floor, electrical hazards are a constant presence. The first step in developing an electrical safety program is to identify the hazards that exist in the facility. Hazards of electricity include electric shock, arc flash, and electrically ignited fires or explosions. Most safety professionals are not electricians or familiar with the many types of electrical hazards. Appendix A provides an overview of the fundamentals of electricity and electrical hazards.

Electric shock can occur from touching or getting too close to current-carrying parts or parts with static electricity. Electric shock can also occur by being struck by lightning or by touching conductive objects that are struck by lightning. Shock is caused by electrical charge passing through the body. Arc flash is caused by electrical charge passing through an object that cannot withstand the energy surge. A metal object or a portion of it can vaporize and create a rapidly expanding ball of hot, gaseous metal and metal fragments. Getting in the way of an arc flash can result in severe burns and even death. Fire and explosion are not usually thought of as electrical hazards, but electricity is a well-recognized source of ignition. Wherever combustible materials are stored, discarded, or allowed to accumulate around hot electrical equipment (such as motors), there is potential for fire. The prevention of this accident requires good housekeeping and good preventive maintenance of electrical equipment. In areas where there is a reasonable likelihood of a flammable atmosphere, special electrical precautions are required. The design, installation, and preventive maintenance of the electrical installations in hazardous locations are of particular interest to electrical safety programs.

After determining the types of hazards present, try to identify the jobs or tasks that expose personnel to those hazards. Anyone who works with portable or stationary tools powered by electricity is potentially at risk. Other tasks that pose electrical hazards are battery charging and operational testing of electrical equipment, assemblies, or components. Electricians, mechanics that work on electrically powered equipment, and welders are at significant risk of electric shock. Even tasks with no intentional contact with electrical conductors or equipment can pose a risk. Personnel working on raised platforms or ladders (such as painters and roofers) or personnel working with long poles (such as workers near cleaning vessels, pits, and swimming pools) could quickly and unintentionally find themselves in contact with exposed live parts. Supervisors are also at risk if they place themselves in a position to observe closely, inspect, or demonstrate procedures to personnel exposed to electrical hazards.

The combination of specific hazards, their location, voltage, and other electrical parameters, and the jobs or tasks that expose personnel to those hazards define the scope of the electrical safety problem. The electrical safety program is an attempt to minimize the risks present within that scope. Prohibiting certain tasks for which there are inadequate safeguards, training, equipment, or knowledge to accomplish the task safely can eliminate some risk. Such decisions are part of the overall electrical safety strategy and program.

Components of an Electrical Safety Program

The electrical safety program can consist of separate policies, programs, and procedures or may take the form of a comprehensive document. Whether the electrical safety program is written under that title or incorporated within other programs is not important. What is important is that the essential elements of an electrical safety program are adequately addressed.

An electrical safety program is a compilation of management practices that minimize the risk of injury from electric shock, flash burns, and electrically ignited fires and explosions. A good electrical safety program builds on the knowledge and experience of the personnel primarily responsible for the care and maintenance of electrical distribution systems and equipment. An electrical safety program should include these essential elements:

- A strategy to manage electrical safety
- Job Hazard Analysis and electrical PPE
- Lockout/tagout of electrical systems and powered equipment
- Procedures for working on exposed energized parts
- Electrical safety training
- Safety management review of new and modified installations
- Periodic safety inspection
- Preventive maintenance

These essential elements are discussed in more detail in the following chapters. Note that no mention is made of installation methods and safety rules (unless those rules fall under safe work procedures). An electrical safety program is not a reiteration of the OSHA regulations or the NEC. The requirements and rules are not a substitute for a methodology to manage the risk associated with operating and maintaining electrical systems. Furthermore, a good electrical safety program will follow the dynamic model for a safety management system discussed in Chapter 1.

To develop an effective electrical safety program, follow this simple outline:

1. Carefully consider the types of electrical hazards, their magnitudes, and the qualifications of personnel exposed to those hazards. This may be accomplished by performing the hazard identification portion of a Job Hazard Analysis (see Chapter 3).

2. Once the potential electrical hazards and risks are identified, the development of the program can proceed by formulating an overall strategy for how to manage that risk (See the discussion of electrical safety strategies later in this chapter).

3. With the chosen electrical safety strategy as a guide, complete the Job Hazard Analysis by determining appropriate controls for each hazard identified (see Chapter 3).

4. Ensure that an adequate means for lockout exists and that each machine has an adequate lockout procedure (see Chapter 4).

5. Develop safe work procedures for planning and accomplishing non-routine electrical maintenance tasks (see Chapter 4).

6. Where hazards cannot be eliminated through lockout/tagout, identify needed electrical personal protective equipment (see Chapter 3).

7. Conduct training for all personnel on the existence and use of these procedures (see Chapter 5).

8. Develop a safety management review process whereby new and significantly modified electrical installations and equipment are reviewed for compliance with federal, state, and local requirements and good safety practices (see Chapter 6).

9. Develop a safety inspection program for the facility (see Chapter 7). The program should require all levels of management to get involved. The safety professional, union safety representatives, and the safety committee (if they exist) should also be involved. Develop a checklist to be used. Several different checklists could be developed depending on the amount of time required for inspection, areas to be covered, level of detail, and knowledge of inspectors.

10. Finally, work with the maintenance manager to develop a preventive maintenance program to ensure that safe installations complying with the NEC remain in good and safe condition (see Chapter 7).

Choosing an Electrical Safety Strategy

A strategy to manage risk posed by electrical hazards is the foundation of an electrical safety program. The strategy is the rule that management uses to determine the level of risk that is acceptable. A strategy can be well conceived or unintentional. In the latter case, the strategy is often that electrical safety is at the discretion of the maintenance personnel. The trouble with that strategy is that management is unaware of the level of risk that they are taking. Therefore, the best strategy is that which facilitates management to maintain risk at an acceptable level. Part of a strategy may be to limit the voltage and types of equipment that employees may work on. Or the strategy might include limitations on the types of tasks that qualified employees can undertake. The reason for such limitations may be an inadequate knowledge base, inadequate tools or personal protective equipment, or inherent risks that management finds unacceptable regardless of the controls. Once the overall strategy is formed, job-specific rules can be developed to minimize the consequences of electrical hazards. Management structure, level of supervision, and the level of resident expertise affect the choice of strategy.

Factors Affecting the Choice of Strategy

Where qualified employees are part of a maintenance department under a single manager, mutual cooperation and support among the qualified employees is expected. A lack of knowledge or experience on the part of one qualified employee can be handled by assigning a team of personnel to a project or by providing a mentor to show how the task is supposed to be done. However, a management structure with a single maintenance department does not always exist. Where qualified employees provide sole support to a production department or area, less cooperation and support can be expected between different departments. They report and take direction from different supervisors and managers. This becomes an impediment to arranging a team of qualified employees from different areas or to assigning a mentor from one area to assist in a different area. Therefore, the management structure within which the qualified employees function may limit the types of tasks that can be undertaken. Not all tasks can be safely accomplished in all departments because some employees lack experience. If assistance is not forthcoming, certain tasks cannot be safely accomplished in some departments. This situation may lead to allowing some tasks to be undertaken by certain qualified employees and not by others. Some method of tracking who is authorized to perform high-risk tasks and who is not is therefore required. When the administration of such a strategy is too difficult, the facility will often decide to prohibit certain high-risk tasks throughout the facility.

Where the level of supervision is low, certain non-routine tasks may not be safe to undertake because of a lack of planning. Non-routine tasks require proper planning to determine the hazards and needed controls so the residual risk will be acceptable. Without supervision to facilitate such planning and perform risk management, non-routine tasks continue to present an unknown amount of risk for the company. This risk can be extreme, leading to death or catastrophe. Where the level of supervision is high, non-routine tasks can be undertaken with due care. Depending on the level of supervision provided in the organization, the strategy might prohibit certain high-risk tasks, especially non-routine tasks.

Where the level expertise of qualified employees is high, management can be fairly confident in giving them tasks with significant risk to perform. Where the level of expertise is not high, certain tasks will be prohibitively risky. It is essential that management determine what types of tasks are too risky. The best way to determine what tasks are beyond the capability of a group of qualified employees is to ask them about their capabilities, what types of equipment would be needed to perform the task, what experience they have had performing those tasks. The qualified employees should be able to help management determine what tasks should not be attempted.

Premises Wiring or Power Generation, Transmission, and Distribution

The first step in choosing an electrical safety strategy is to identify the electrical hazards present. This can take the form of the first part of a Job Hazard Analysis (JHA), hazard identification (see the discussion of hazard identification in Chapter 1). In determining the electrical hazards present, first determine the line of demarcation between the facility's responsibility and the utility's responsibility. Then, determine what part of the facility's electrical installation is premises wiring and what part, if any, is considered power generation, transmission, and distribution.

Federal OSHA regulations make a distinction between premises wiring and electric power generation, transmission, and distribution. Federal safety regulations for premises wiring are contained in Title 29, *Code of Federal Regulations*, Part 1910, Subpart S (29 CFR 1910 Subpart S). However, in most cases throughout the United States, state and local safety regulations for premises wiring are contained in the NEC. Municipalities and, in some cases, states adopt a particular edition of the NEC to enforce. Since the NEC adopted by the state or local municipality is usually much more recent than the NEC from which OSHA's Subpart S was developed, the state or local code usually contains the more stringent requirements.

Federal safety regulations for electric power generation, transmission, and distribution are contained in 29 CFR 1910.269. Municipalities and/or states usually adopt a particular edition of the NESC. The federal regulations were developed from the NESC. Similar to the regulations for premises wiring, the state or local code for electric power generation, transmission, and distribution usually contains the more stringent requirements.

The line of demarcation between electrical power generation and distribution systems and premises wiring is not always clear. When there is no power generation equipment in the facility, the premises wiring starts at the service point identified by the electric utility that provides power to the facility. Typically, the service point is defined by the utility as the utilization side of the meter on a service lateral, or where the overhead service drop connects to the building, but sometimes at the transformer on a pole.

Standby and emergency power equipment is considered premises wiring and is covered by the NEC and OSHA's Subpart S.

Where there is electric power generation equipment for other than standby or emergency power, the generating equipment will be subject to the NESC and OSHA's 29 CFR 1910.269 while the premises wiring will be subject to the NEC and OSHA's Subpart S. However, the employer has some latitude in deciding the line of demarcation between the two. In determining the line of demarcation, it is essential to provide consistent, clear instruction to maintenance personnel so that there is no ambiguity or confusion. Avoid treating the same type of equipment in two different ways within the facility (e.g., bus bar in the plant with vertical drops to equipment or outlets treated as premises wiring in some instances and as power distribution system subject to 29 CFR 1910.269 in other cases). For example, overhead power lines between a powerhouse and a factory building may be at two different voltages, but would consist of identical equipment (see Figure 2.1). If the service point is on the secondary side of the utility transformer, then the overhead lines from the pole to the transformer are not the responsibility of the factory employer. Therefore, the remaining overhead lines could be considered premises wiring. The generator and switchgear including the service disconnect, generator disconnect, and main breaker to the factory building – all located in the powerhouse – could be considered power generation, transmission, and distribution equipment. This would create a clear demarcation between power generation equipment and premises wiring. However, if the employer determines that the high side of the step-down transformer is to be considered power distribution equipment, then one set of overhead lines is power transmission equipment while the other set of overhead lines (back to the powerhouse auxiliary loads) is premises wiring. Furthermore, a second demarcation line is required to separate the high voltage motor load (premises

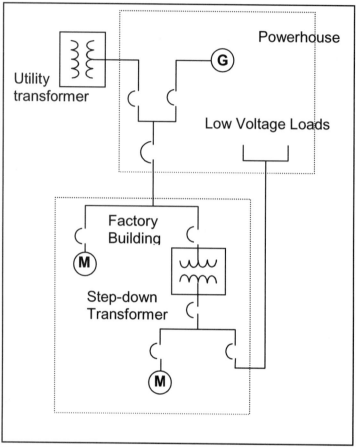

Figure 2.1: One-Line Diagram of Factory

wiring) from the high voltage bus feeding the step-down transformer. This situation is fraught with confusion and danger, and should be avoided.

Understanding the Risk Posed by Electrical Hazards

The next step to choosing the appropriate risk management strategy is to identify the actual risks posed by electrical installations and equipment at your facility. There are two components of risk: severity and exposure (see Chapter 1). It is not necessary to perform a formal hazard analysis at this time. What is important is to identify the voltages at which equipment in your facility operates and under what circumstances personnel could be exposed to live parts.

The best resource to assist in this endeavor is the personnel who are primarily responsible for the care and maintenance of the electrical distribution system and electrical equipment. Their knowledge and experience should be tapped by utilizing either a team approach to the development of the electrical safety program or by delegation of specific action items to appropriate personnel with the requisite knowledge and experience. Either management approach can work effectively depending on the work culture at the facility.

One starting place, and a really useful tool, is a one-line diagram of the electrical distribution system throughout the facility. Such a one-line diagram consists of identification of the service point or points and the distribution of electricity throughout the facility. The voltages at each point along the distribution are identified on the diagram. The diagram also identifies all overcurrent protection devices and disconnecting means. On this diagram it is a good idea to mark the boundary between equipment that is the primary responsibility of the utility provider and equipment that is the primary responsibility of the owner of the facility. If this diagram exists, the number of service points, the voltages provided to the facility, and the scope of electrical installations that are the responsibility of the facility owner are easily identified. If this diagram does not exist, its development should be begun so that this essential information is documented for future use.

In addition to the electrical distribution system, specific pieces of equipment may use rectifiers and transformers to convert electricity to direct current (DC), high voltage, or low voltage applications within the machine. All high voltage applications and the associated equipment should be identified and documented at this time. If machine-specific lockout procedures have been developed, the procedures may contain some of this information. Labels on electrical equipment are another good source for voltage, phase, and type information. And manufacturer's specifications and instruction manuals can also provide electrical rating information.

In addition to high voltage (>50 V) AC and DC motive equipment, electrostatic generating equipment may be hazardous depending on the amount of charge stored.

Once the voltages of the distribution system and equipment are known, the next step is to identify the personnel potentially exposed to these voltages. Typically, operators of electromotive equipment are not exposed to electrical hazards because the live circuit parts are insulated, enclosed, or otherwise guarded. However, this is not always the case, and assuming a lack of exposure is dangerous. If operators are permitted to open electrical cabinets and panels for any reason, then there is exposure. Cleaning and painting are tasks with potential exposure to live electrical parts. Usually, the maintenance personnel with primary responsibility for the care and maintenance of facilities and machines have the greatest potential for exposure to electrical hazards.

For each department, job classification, and type of electrical enclosure or equipment, determine:

1. Who is required to have access to exposed live parts

2. Why work will be performed by those personnel within the limited approach boundary

3. How often is such access required

4. What level of training is required for personnel to be competent to perform that work

5. What personal protective equipment would be necessary for personnel to perform that work safely

Setting Policies

Now that the hazards and the personnel likely to be exposed have been identified, consider some safety policies. Policies must be well considered, communicated, and enforced. Policies must consider the reduction of risk, the impact the policy will have on production and maintenance efforts, the impact on training and equipment that must be provided, the ease of enforcement, and the burden of enforcement on supervisors, managers, and support staff (i.e., safety department). Examples of pertinent electrical safety policies could be:

- No work within the restricted approach boundary will be performed on voltages above 480 VAC.

- All work other than voltage measurement within the restricted approach boundary must be approved by the Facilities Manager.

- All work other within the restricted approach boundary requires the use of electrical insulating PPE and arc flash PPE appropriate for the voltage unless specifically approved by the Maintenance Shift Supervisor.

- All riggers, painters, and programmable logic controller programmers must receive electrical safety training at least at the non-qualified employee level.

- No employee is permitted to approach the 1200 VAC circuit in the Mill Department within the restricted boundary unless it is completely de-energized and locked out. All live work on this circuit will be performed by a qualified contractor.

Safety policies will be further developed into safety rules and safe work procedures. All policies, rules, and procedures must be incorporated into training. Safety training must be appropriate for each job classification and must be communicated clearly and frequently to ensure understanding and retention of knowledge. Compliance with policies, rules, and procedures must be monitored and enforced within the limitations of contractual agreements. Enforcement efforts should not get bogged down in legalism, but lack of enforcement sends a dangerous message and sets a tone that negates all efforts to manage risk.

Table 2.1: Height of Exposed Live Parts above Work Platform or Floor

50 – 600 V	8 feet
601 – 7,500 V	8.5 feet
7501 – 35,000 V	9 feet
> 35, 000 V	9 feet + 0.37 inches per kV above 35 kV

Controlling Access to Live Parts

Live parts must be enclosed or safeguarded wherever possible. Overhead lines and crane buses must be safeguarded by height above and distance from all work platforms. Exposed live parts must be at least at the heights indicated in Table 2.1.

High voltage parts must be placed in rooms or vaults where access is restricted to authorized personnel only. All electrical enclosures must be complete (i.e., without holes or openings). All personnel who are not qualified to work on or near live parts must be made aware that they are not permitted to open any electrical enclosure where live parts are exposed. It is essential that systems are designed such that unqualified personnel do not have to open, for any reason, electrical cabinets where live parts are exposed. All indicators and controls must be on the outside of electrical cabinets. Where exposed parts are safeguarded by height above the work platform, they must be clearly identified as hazardous. Personnel who work on ladders and mobile work platforms and who operate high lift powered industrial trucks must be taught the location and hazards of exposed live parts safeguarded by height. These personnel may need to be trained as qualified to work near exposed live parts. Maintenance and engineering personnel must not leave electrical cabinets open so that live parts are exposed. If access to live parts cannot be limited to qualified personnel by these means, then electrical enclosures must be locked to prevent unauthorized access. Consider that electrical cabinets in places frequented by non-employees such as in waiting rooms or reception areas should always be locked to prevent unauthorized access.

Limiting Voltage for Live Work

First, determine the maximum voltage that personnel will be qualified to work on. For most industrial facilities, a maximum voltage of 480 VAC or 600 VAC is sufficient to allow qualified employees to troubleshoot most machinery and electrical installations. However, the voltage limitation will not allow employees to open the main transformer cabinet at the service point (if on the premises). The reason for the voltage limitation policy is to protect employees who may not have the training or equipment to work on or near such high voltages. Many companies choose to hire outside contractors to perform work in this cabinet rather than risk their own employees. Through such a policy, the companies limit the training and personal protective equipment that they must provide.

Requiring Authorization for Live Work

Second, state explicitly that circuits to be worked on are to be de-energized whenever possible. This policy addresses the intent of electrical safety regulations. Third, determine who can authorize work on live parts, and under what conditions. For example, the Plant Manager, Maintenance Manager, Plant Engineer, or Safety Manager are possible choices for delegating the authority to approve work on live parts. Conditions that warrant live work include troubleshooting or diagnostic work, work on low voltage (less than 50 V), or where de-energizing the circuit would cause a greater hazard. The latter condition is often incorporated into the regulations of healthcare facilities where lives depend on electrical power and circuits cannot always be de-energized to work on them. This condition is not generally applicable to industrial facilities. By going on record, management makes it clear to first-line supervisors and employees alike that working on live circuits for mere convenience or expediency will not be allowed. Such a policy demonstrates management is committed to modifying equipment and electrical installations as necessary to eliminate any perceived need to work on live parts.

Developing Qualifications

Where risky tasks must be performed, personnel must be competent, qualified, trained, equipped, and supervised. A qualified employee is one who the employer has full confidence in to perform work on or near exposed live parts of electrical circuits. The regulations and standards require that a qualified employee be competent to perform their work and receive specific training. The federal and state regulations and industry consensus standards do not dictate who can and cannot be qualified. OSHA does not require that qualified employees be licensed electricians or journeymen. It is up to the employer to determine who is qualified and who is not. And OSHA holds the employer responsible for the safety of those qualified employees.

> **Local ordinances often require a facility's electrical installation to be performed by a licensed electrician under a builder's permit. To determine what types of new and modified installations require a permit and special license, and what types of electrical work do not, check your local authority having jurisdiction.**

Qualified employees must be competent and trained. They must be competent through training, education, or experience to work on or near exposed live parts. The employer is responsible to ensure an employee is competent to perform the assigned task. Therefore, the safety professional, maintenance manager, and senior facility manager should determine what qualifications are needed for the various jobs that require working on or near exposed live parts. They should work with the Human Resources department to include these qualifications in the job description and ensure that needed training topics are developed and offered to employees in order for them to be considered qualified. This will ensure that new hires have the skills they need to be qualified at the facility and that qualified employees are adequately trained. Qualifications that would be indicative of competence to work on or near exposed live parts are dependent on the type of job.

A licensed electrician would certainly be competent to perform almost any electrical maintenance, installation, or repair. The only limitation would be his/her experience (for example, prior work on 6,200 V equipment). However, a license is not usually required. If a person is going to perform electrical maintenance or repair, he/she should certainly know electrical terms, how to measure voltage, how to solder, and how to connect wires to terminal screws. Such knowledge may come from formal training or personal experience. Prior work experience in electrical maintenance with a good recommendation from his/her previous employer would indicate general competence. Additionally, the ability to read and write in English and to speak English might be necessary if the programs, policies, and procedures are in English. Supervisors may not speak any language other than English. Equipment operating and maintenance manuals and troubleshooting guides are typically in English, although many are available in Spanish. Equipment purchased from European countries often come with English, German, and French text. Therefore, consider the language requirements of the job.

In addition to being competent for the position, all qualified employees must receive training beyond that required for unqualified employees. Specifically, qualified employees must be trained on the following:

- Distinguishing between live parts and other parts
- Safe approach distances to exposed live parts
- Voltage measurement
- Hazard determination and job planning process
- Personal protective equipment
- Visual inspection of test instruments and equipment
- Notification and disposition of damaged or defective test instruments and equipment
- Instrument and equipment ratings
- Special precautionary techniques
- Insulating and shielding materials
- Insulated tools

The employer provides this training to qualified employees, and the training must be specific to the hazards, procedures, and policies at the facility. Generic training, video, interactive, or online training is not sufficient by itself. Electrical safety training is discussed further in Chapter 5. The qualified employee has the knowledge and skills to perform the assigned tasks, but how does the employer go about ensuring continued competence?

Verifying Competency

Regardless of a person's training, education, and past work history, the employer must verify a qualified person is competent to perform the assigned tasks. This is true for new hires and more established employees as well. Ensuring competence should incorporate periodic review (annually if possible) and continuing evaluation. The periodic review should be a formal approach, such as a performance review. And it can incorporate observation of work practices. This periodic review of the competence of a qualified employee can and should be integrated with elements of the periodic review of the lockout/tagout program required by OSHA. Lockout/tagout will be discussed further in Chapter 4. The periodic review of lockout/tagout requires discussing the duties and responsibilities of an employee who is authorized to conduct lockout/tagout. The lockout/tagout periodic review also requires observing authorized employees performing lockout/tagout in accordance with machine-specific procedures. Since most qualified employees will also be authorized employees under the lockout/tagout program, this review will show knowledge retention and safe work practices needed to work on de-energized electrical parts. As for safe work practices involving exposed live electrical parts, the employer should formally evaluate each qualified employee once a year. But the qualified employee competency review should also be integrated with programs that provide opportunities for evaluation throughout the year.

Many opportunities exist to verify the competency of a qualified employee throughout the year. Such opportunities include updating a job hazard analysis, investigating an accident or incident, performing a behavior-based safety observation, conducting a safety inspection or walkthrough, or inspecting the status or quality of a job. The results of these activities should be documented, reviewed by the supervisor, and communicated back to the qualified employee. When safe practices, compliance with electrical safety rules and procedures, good judgement, and quality workmanship are observed, the qualified employee is found competent. Where deficiencies are noted, counseling or additional training may be needed. Even disciplinary action may be required. Serious lapses in judgement, disregard for safety, or habitually poor workmanship would indicate a lack of competence. If an employee shows a lack of competence at any time, the employer should remove him/her from the list of qualified employees, change the assigned tasks accordingly, and take the steps necessary to ensure this person no longer performs work on or near exposed live parts. If the employer has reason to doubt an employee's competence and does not take prudent action, the employer could be found negligent in the case of an accident. The need for verifying competence and the degree of competence and confidence the employer affords a qualified employee determines the level of supervision required.

Providing Supervision

No matter how well trained and competent employees are, they must be held accountable for their actions. Accountability is an essential part of the management process. The question is how much supervision is needed and when is it needed. To answer this question, the supervisor needs to be familiar with the employee and the job. In the following discussion, the term supervisor means the member of the management team responsible for the work of qualified employees or the supervisor's designees.

The supervisor must be familiar with the employee's knowledge, skill, and experience. And the supervisor must be familiar with the demands of the assigned task, its level of difficulty, the hazards, and the necessary controls. The better the match between the qualified employee and the assigned task, the less supervision is required. However, some supervision is always required. Without adequate supervision, there is no accountability. Without accountability, the management process is incomplete. Consider that supervision provides:

- Compliance with policies and procedures

- Assurance of high quality work

- A learning/teaching opportunity (sometimes for both the worker and the supervisor)

- Management demonstration of the importance of the work

- An opportunity to praise performance

A common perception is that supervision is about compliance and quality assurance. The last three opportunities in the list above, however, are what management should strive for and be trained to accomplish.

An opportunity to see what an employee is capable of and to provide additional on-the-job training enhances the worker's skill and knowledge as well as the mentor's teaching abilities. Anyone with more knowledge or skill applicable to the assigned task than the worker can teach or mentor. Supervisors, shift leads, peers, or even contractors can be effective mentors and teachers. When a supervisor observes or checks up on the work, it demonstrates the importance of the work. If this is done with sensitivity to the feelings of the worker, it can build self-esteem and pride of workmanship. The Hawthorne Effect is a well-known and established relationship between management attention and worker job satisfaction. The more management attention given to work conditions, the higher the worker's job satisfaction and productivity. This is also true of any genuine demonstration by management that the worker performs an important function for the company. When supervision is combined with mentoring and sincere praise of good performance, great management/employee relations can result—a sense of teamwork, pride of workmanship, pride in the company, better compliance with policies and procedures, and safer workplace behavior.

Supervision also affords the supervisor with the opportunity to learn more about the task being performed and about the worker. It can be an opportunity for the supervisor to study a task in greater detail than he/she has ever done before. It can provide needed insight into the difficulty or impossibility of following procedures verbatim, of inadequate job hazard analyses, of accepted unsafe practices that can be eliminated or modified to reduce unnecessary risk.

With all the responsibilities a supervisor has, the primary function (i.e., to supervise the workers) often takes a backseat to other more pressing matters, like paperwork. All too often, the supervisor spends so much time attending to the most recent "issue" that there is no time left to adequately supervise the workers. Adequate supervision means accomplishing all five of the functions listed above, not just the first two. One way to accomplish supervision is for the supervisor to delegate some of this responsibility to senior employees who have the skills and desire to perform supervisory functions. But there is no substitute for the supervisor's periodic and personal involvement in demonstrating that management cares about the work and the worker. As previously stated, the supervisor can also benefit from such encounters. Delegation of supervisory tasks is good, but the supervisor must hold employees accountable for their performance. There is no better way for the supervisor to determine performance than by direct observation.

Chapter Three

Job Hazard Analysis and Electrical PPE

Overview

Each job or task on or near exposed live electrical parts should be analyzed to identify the types of electrical hazards, their magnitude, and the qualifications of the personnel exposed to those hazards. Routine and non-routine tasks should be considered. OSHA regulations require a personal protective equipment (PPE) assessment for all tasks. The Job Hazard Analysis (JHA) is the preferred method for accomplishing the PPE assessment. The JHA is an analysis method by which the hazards associated with a particular task are identified. The JHA seeks to determine if the hazard controls are adequate to reduce the risk to an acceptable level. The JHA is frequently used to identify physical and chemical hazards and ergonomic risk factors.

JHA consists of identification of hazards and determination of controls to minimize the risk of those hazards. All hazards, not just electrical hazards, are included in the analysis. All types of controls, not just PPE (the least preferred method), should be considered. The hazard identification of the JHA is the basis for forming the electrical safety management strategy. However, the safety management strategy is the basis for determining the types of controls to be used. Therefore, the JHA and the electrical safety management strategy go hand-in-hand. Once controls have been selected for the various tasks, an electrical PPE program can be created or incorporated into the general PPE program.

Federal OSHA regulations on PPE are contained in [PPE] General Requirements (29 CFR 1910.132), Electrical Protective Devices (29 CFR 1910.137), Safeguards for Personnel Protection (29 CFR 1910.335), and Electric Power Generation, Transmission, and Distribution (29 CFR 1910.269). OSHAs Subpart S was partially derived from the Standard for Electrical Safety Requirements for Employee Workplaces (NFPA 70E). NFPA 70E – 2000, Part II, Section 3 contains the most current information concerning electrical PPE, especially for protection from arc flash. See the discussion on selecting electrical PPE later in this chapter.

It is more efficient to incorporate tasks with electrical hazards into the existing JHA or to perform a single JHA for all tasks than to initiate separate programs or analysis efforts. The efficiency comes as JHAs are updated, modified, and revamped. The nature of business today is continuous change and improvement. One JHA process can be managed better than multiple separate efforts.

How to Conduct a JHA

The first thing to do is to determine what the JHA is going to accomplish. How will the results of the JHA be used. For example, if the JHA is a PPE assessment only, the evaluator does not need to be familiar with machine safeguarding requirements. If the JHA will be used to determine risk, the field form should contain a place for the evaluator to document the potential severity and likelihood of the hazards resulting in an accident. After the goals of the JHA are determined, develop a field form so that evaluations will be consistent. A field form can be taken on a clipboard and filled out while at the job site to document the evaluation of a specific job. The form should help prevent the evaluators from inadvertently omitting anything significant. A checklist is often much more effective at preventing omissions than a fill-in-the-blank form. The checklist could incorporate interview questions that the evaluator should ask operators or themselves when analyzing a specific job. OSHA 3071 provides the following list of questions as examples of the kinds of questions that could be asked:

- Are there materials on the floor that could trip a worker?
- Is lighting adequate?
- Are there any live electrical hazards at the job site?
- Are there any chemical, physical, biological, or radiation hazards associated with the job or likely to develop?
- Are tools—including hand tools, machines, and equipment—in need of repair?
- Is there excessive noise in the work area, hindering worker communication or causing hearing loss?
- Are job procedures known and are they followed or modified?
- Are emergency exits clearly marked?
- Are trucks or motorized vehicles properly equipped with brakes, overhead guards, backup signals, horns, steering gear, and identification as necessary?
- Are employees operating vehicles and equipment properly trained and authorized?
- Are employees wearing proper PPE for the jobs they are performing?
- Have any employees complained of headaches, breathing problems, dizziness, or strong odors?
- Is ventilation adequate, especially in confined or enclosed spaces?
- Have tests been made for oxygen deficiency and toxic fumes in confined spaces before entry?
- Are workstations and tools designed to prevent back and wrist injuries?
- Are employees trained in the event of a fire, explosion, or toxic gas release?

Obviously, this list is not all-inclusive, nor is it necessary to include every question in this list on the field form. The development of the form is a good place to involve employees, especially those who may be called upon to conduct the JHA evaluations.

The example in Figure 3.1 of a JHA field form was developed for a company that did not require a risk determination. The form is in two sections and was copied double sided so that there were two pieces of paper for each task. The two pieces of paper can be stapled together before the evaluation. That way, the description of the task, department, location, and other identifying information only needs to be filled out on the first page.

After the form is developed, the employees who will conduct the JHA should be selected from among interested and conscientious employees, supervisors, managers, safety professionals, and other resources such as temporary employees and consultants. These personnel must be trained. The training should include:

- A description of a JHA
- Why a JHA is being done
- Familiarization with the field form
- The procedure to follow
- Who to contact with questions or problems

It is a good idea to conduct a walkthrough of a job evaluation allowing everyone on the team to fill out a field form. The results should be compared and discussed. This should be repeated as many times as necessary for the members of the team and management to feel comfortable and confident that the JHA will be consistent, accurate, and complete. Then, responsibilities for various jobs, departments, or areas can be divided and assigned to individuals for completion.

The conduct of the JHA itself is very simple. The evaluator should introduce him/herself to the worker and explain that they are going to observe the job to identify any hazards and their controls. Tell the worker that his/her name will not be documented, and that this analysis is not about performance or productivity. Ask the worker the interview questions on the form and any other questions that will: make the worker more comfortable; help the evaluator understand what the job entails and what the demands of the job are; and encourage the worker to talk openly about the issues of the job and the work area.

Next, observe the job and identify the hazards. If the job consists of tasks or steps that have different sets of hazards and there is delay or preparation time between tasks, then break the job into components. If the hazards are consistent throughout all portions of the job, or there is no time between tasks, treat the job as a whole. Document the hazards identified with the job or each of its tasks. For every hazard, identify what control is present that minimizes its risk. Where the hazards are inadequately controlled, make recommendations for control. If the evaluator cannot think of a good recommendation, he/she should document that the controls are inadequate. Recommendations can be added later. If PPE is needed, complete the PPE assessment portion of the field form. This form allows the evaluator to specify PPE for the job or for separate tasks within the job.

After the job is documented, the safety professional should review the field forms for completeness and sound safety principles. If there is missing information, the field form should be returned to

COMPANY - LOCATION
PERSONAL PROTECTIVE EQUIPMENT AND JOB SAFETY ASSESSMENT

Department: _____   Job Code: _____   Task: _____
Location: _____   _____   _____
BT#: _____   _____   _____
Equip Name: _____   _____   _____

EYE/FACE HAZARDS			PPE REQUIRED	
Impact (Large Fragments)			Safety glasses w / sideshields	
Heat	Hot Sparks		Faceshield / goggles / safety glasses	
	Molten metal splash		Faceshield / goggles	
	High temperature		Screen/reflective faceshield	
Chemical	Splash (Specify)		Goggles & faceshield	
Exposure	Mist or Vapor		Goggles (indirectly vented)	
Dust			Goggles	
Optical Radiation	Arc Welding		Welding Helmet	
			Shade 10-14 Shade *	
	Gas Welding		Goggles / Helmet Shade 4-8	
	Cutting/Brazing		Goggles / faceshield Shade 3-6	
	Glare		Glare-reducing spectacles	
	Laser - Class		Laser goggles	
Other				
Comment				

* Rod manufacturer's minimum recommended shade

HEAD HAZARDS			PPE REQUIRED	
Impact	Falling objects		Helmet	
	Swinging/Rotating Objects		Helmet	
	Bumps		Bump cap	
Heat	Sparks		Welder's cap	
	Molten metal / high temperature		Helmet	
Chemical Splash/Mist (Specify)			Tyvek head covering	
Electrical - voltage V			≥ 2,200 V ≤ 2,200 V	
Other				
Comment				

HAND/ARM HAZARDS			PPE REQUIRED	
Heat	Thermal Bums		Hot Mill/Aluminized Gloves	
	Welding/Sparks		Welder's Glove	
Laceration/Abrasion/Pinch			Leather / heavy cotton glove	
Puncture			Kevlar glove	
Electrical - voltage: _____ V			Electrical glove / sleeves	
Vibration			Anti-Vibration Gloves	
Cryogenic			Cryogenic Gloves & sleeves	
Chemical Exposure (Specify)			_____ gloves / barrier cream	
UV Radiation			Opaque shirt & sleeves / gloves / sunscreen	
Other				
Comment				

Figure 3.1: Sample JHA Field Form

TORSO/LEG HAZARDS			PPE REQUIRED	
Heat	Thermal burns/molten metal splash		Reflective clothing	
	Welding/Sparks		Welding Jacket/ Long pants	
	Fire		Fire-resistant clothing (e.g., Nomex)	
Chemical Exposure (Specify)			Chemical Resistant Clothing (specify type)	
Electricity- voltage V			Insulating blankets / mats	
Cuts/Abrasions			Leather Apron/knee pads	
Cryogenic			Insulated coverall	
UV Radiation			Opaque shirt / long pants / sunscreen	
Other				
Comment				

FOOT HAZARDS		PPE REQUIRED	
Impact		Safety shoe - 75/50/30 ft. lb.	
Compression		Safety shoe 2500/1750/1000 lb.	
Sole Puncture		Safety shoe - PR	
Static Electricity (Semiconductor)		Safety shoe	
Voltage V			
Electricity - voltage V		Safety shoe - EH	
Conductivity - voltage V		Safety shoe	
Metatarsal damage		Metatarsal cover - 75/50/30 ft. -lb.	
Chemical exposure (Specify)		Chemical-resistant boot	
Slippery surfaces		Slip-resistant soles	
Molten metal/extreme heat		Aluminized spats	
Other			
Comment			

ENTIRE BODY HAZARDS		PPE REQUIRED	
Fall		Fall Restraint - Belt/Lanyard	
Fall		Full Arrest Harness with Lifeline	
Cold wealher		Cold weather gear	
Wet weather		Wet weather gear	
Other			
Comment			

FID#	CHEMICAL NAME	HAZARDOUS COMPONENT	RESPIRATORY HAZARDS

RESPIRATOR NEEDED: _____

NOISE HAZARDS		PPE REQUIRED	
Sound Level dB A		Hearing Protection	

Comments:
_____
_____

Evaluator's Initials: _____   Date: _____

Figure 3.1: Sample JHA Field Form (cont.)

COMPANY - LOCATION
PERSONAL PROTECTIVE EQUIPMENT AND JOB SAFETY ASSESSMENT

Department: _____   Job Code: _____   Task: _____
Bay Location: _____   _____   _____
BT#: _____   _____   _____
Equip Name: _____   _____   _____

WORK AREA HAZARDS		CONTROLS
Means of egress		
Illumination		
Adequate ventilation		
Truck traffic		
Tripping		
Slipping		
Falling through opening		

TASK	HAZARD	CONTROLS

Figure 3.1: Sample JHA Field Form (cont.)

TASK	HAZARD	CONTROLS

PROCEDURES

Operating procedures	
Emergency procedures	
Lockout procedures	
Confined space entry procedures	

ILLNESS/INJURY HISTORY		SUMMARY
Accidents		
Near misses or close calls		
Complaints of headaches, breathing problems, dizziness, or strong odors		

Figure 3.1: Sample JHA Field Form (cont.)

the evaluator for completion. If sound safety principles are not being followed in identifying hazards, controls, or making recommendations, the safety professional may have to provide additional training for the evaluator. Chronic concerns with consistency, accuracy, completeness, or soundness should result in removal of the evaluator from the team conducting the JHA.

The results of the JHA should be typed for distribution and communication to the affected employees. Whereas the checklist format is beneficial in preventing omissions, the same format makes for a very long and superfluous report. Only the pertinent information should be included in the report. Portions of the field form that were not applicable to the job and blank fields, lines, or sections should be omitted from the report. Two examples are provided in Figure 3.2.

COMPANY X – LOCATION Z
JSA & PPE Evaluation

Department: Maintenance
Location: All
Job: Electrical Maintenance

TASK	HAZARD	CONTROL
Troubleshooting (440 V maximum)	Electric shock, arc flash burn	Wear appropriate flash protection clothing and safety glasses when within 3 feet of exposed live parts. Provide Class 00 gloves for use when needed. Recommendation Conduct training on the proper use of electrical PPE.

COMPANY X – LOCATION Z
JSA & PPE Evaluation

Department: Maintenance
Location: All
Job: Welding

TASK	HAZARD	CONTROL
Arc welding and fuel gas welding, cutting and brazing	Eye hazard	Wear a welding helmet with appropriate shade of lenses for welding type. Recommendation Clean dirty lenses and replace broken lenses.
	Burns	Wear welding jacket and gloves Wear welding apron if sparks will strike pants.
	Electric shock	Ensure proper grounding of arc welding electrode. Make sure area is dry.

Figure 3.2: Sample JSA and PPE Evaluations

Hazard/Risk Evaluation for Work on or near Exposed Live Parts

NFPA 70E, Part II, Appendix D, provides an example of a decision-flow diagram for an electrical hazard analysis. This flowchart has been provided in Figure 3.3 to help explain the process of conducting a hazard analysis and to introduce the NFPA 70E standard. It is strongly recommended that the standard be consulted with all of its detailed explanation and requirements.

How to conduct an electrical hazard analysis for a specific task on specific equipment can be accomplished by determining the following:

1. What is the nature of the work to be performed?
2. What are the hazards?
- What hazards are associated with performing that type of work?
- Does performing the work on the particular equipment pose any special hazards?
- Are the conditions under which the work will be done in any way hazardous?
3. What are the safeguards needed to control those hazards to an acceptable level or to the maximum extent feasible?

Step 1

Determine the nature of the work to be performed. The nature of the work will typically fall into one of five categories:

- Electrically non-hazardous work
- Diagnostic work
- Limited work
- Restricted work
- Repair work

These categories are by no means the only ones possible, but they serve as an example of the types of work on or near exposed live parts that may be applicable at an industrial facility.

Step 2

Once the nature of the work is determined, identify hazards associated with that type of work. The hazards of performing the type of work on particular equipment may be further modified or exacerbated by the condition of the equipment at the time the task is performed. Hazards can include: electric shock, arc flash, fire, dangerous equipment fault (e.g., a berserk robot or a boiler without temperature control), fall from height, dropping tools or equipment onto other personnel, entering confined spaces, working around unguarded machinery.

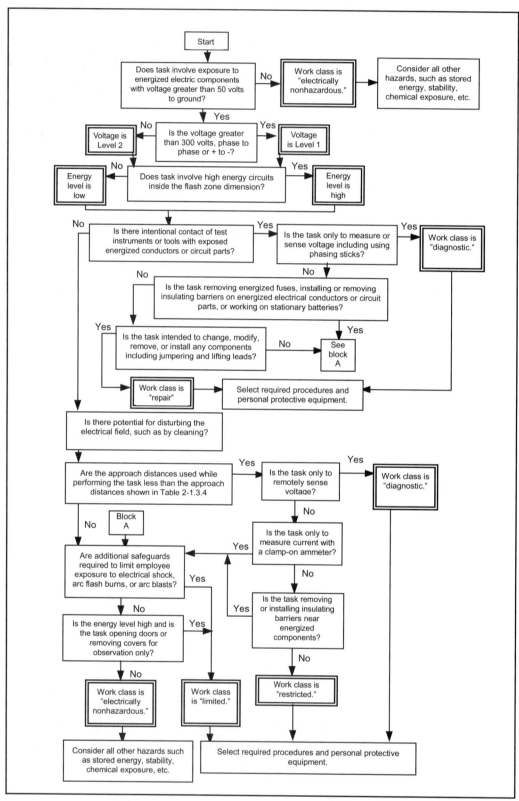

Figure 3.3: NFPA 70E, Part II, Appendix D, Figure D-1 Hazard/Risk Analysis Evaluation Procedure Flow Chart

Step 3

For every hazard identified, a control must be found that will reduce the risk to an acceptable level or to the greatest extent feasible. If the risk remains at an unacceptable level, additional steps should be taken to respond to accidents quickly. Such steps might include additional firefighting equipment and personnel or emergency rescue service personnel standing by.

Electrically Non-Hazardous Work

Electrically non-hazardous work is work on live parts less than 50V or near live parts that are not effectively exposed (i.e., the parts are guarded in some way). There is not any particular electrical hazard associated with this type of work. No safeguards are needed.

Diagnostic Work

Diagnostic work is the measurement of electrical characteristics (e.g., voltage and current) by means of test instruments. This work may be remote or through direct contact with live parts using insulated test devices. This category of work requires simple approach, touch, and withdraw motions using appropriately insulated tools. This type of work is not particularly hazardous for a properly trained qualified employee unless he/she must reach past exposed, energized parts to the test points. A generic procedure for diagnostic work should be developed to guide decisions concerning PPE and other safeguards for diagnostic work.

Limited Work

Limited work is of a routine nature and has relatively low risk. Limited work includes replacing energized fuses with an insulated tool, installing or removing insulating barriers, and attaching clip-on instrument leads. Limited work can also be work performed outside the limited or restricted approach boundary for unqualified or qualified employees, respectively, where some additional safeguard is required to protect the employee. For example, opening covers for observation is limited work when removal of such covers places personnel inside the flash protection boundary. This category of work requires easily accomplished motions while close to exposed live parts. A generic procedure may be developed for limited work, especially for routine tasks such as pulling energized fuses and opening covers.

Restricted Work

Restricted work is work performed inside the approach boundaries that is not diagnostic or limited work. Such work is not routine or poses a greater risk than limited work. Perhaps the conditions are abnormal—such as wet conductors, extreme humidity, or potentially damaged equipment. This category of work requires more difficult or complex motions such as using screwdrivers or wrenches. An administrative procedure describing how to gain approval for such work should be developed. This type of work can be quite hazardous.

Repair Work

Repair work is modifying, removing, or installing components in live electrical circuits. Repair work involves intentionally contacting exposed live parts including installing jumpers and lifting leads. This category of work will always require some type of PPE to protect hands from electric shock. By policy, such work should never be performed if the task can be accomplished with the electrical parts de-energized. An administrative procedure listing the known tasks that fall in this category and how to gain approval for such work should be developed.

Identifying Tasks with Electrical Hazards

JHAs are often performed on production tasks, but rarely are they performed on maintenance tasks. Maintenance tasks are varied and change with circumstances. Therefore, they are more difficult to analyze. However, routine maintenance or service tasks have inherent electrical hazards that can be identified, at least by general classification. Once the tasks are identified and analyzed, specific electrical PPE can be chosen as appropriate to protect employees from those hazards that cannot be eliminated or otherwise controlled.

The first step to incorporate electrical safety into the JHA is to identify specific tasks that have electrical hazards. Such tasks include:

- Battery charging

- Battery care

- Handling portable electric equipment

- Routine opening and closing of circuits

- Arc welding

- Painting or cleaning near exposed live parts

- Working above exposed live parts

- Taking voltage readings on live circuits of various voltages in various types of cabinets (e.g., 120 V programmable logic controllers (PLCs), 480 V switchgear) or equipment

Ask qualified employees what kinds of work they do on live parts. These tasks are not the same as production tasks, so a change in thinking may be necessary for those used to performing JHAs on production tasks. Additionally, NFPA 70E, Part II, Table 3-3.9.1 identifies some routine tasks on various kinds of equipment at various voltages. The list of tasks, included as Table 3.1, is a good place to start. It helps explain the process of determining appropriate PPE and serves as an introduction to the NFPA 70E standard. However, it should not be the only source used to identify tasks with electrical hazards. It is strongly recommended that the NFPA 70E standard be consulted with all of its detailed explanation and requirements.

Identifying Specific Electrical Hazards

Consider the task to be performed, the type of equipment to be serviced, and the environmental conditions surrounding the equipment. Determine the voltages involved in the equipment being worked on. Determine the minimum distance from the exposed live parts that is needed to conduct the task. Determine if the task will be performed within the flash protection boundary, limited approach boundary, restricted approach boundary, or prohibited approach boundary. (See Chapter 4 for a discussion of the approach boundaries.) If working within the flash protection boundary, a flash hazard analysis must be performed to determine the PPE needed to protect employees from arc flash. (See the flash hazard analysis section of Chapter 4.) If work will be performed within the restricted boundary, a careful consideration of the potential to inadvertently cross the prohibited boundary must be made. If the task requires the crossing of the prohibited boundary or if it is credible that the prohibited boundary can be inadvertently crossed by conductive tools, objects, or any part of the body, live parts must be insulated or insulating PPE must be worn.

Determining Adequacy of Controls

The hazard reduction precedence sequence (HRPS) is a term used in system safety to describe the order of preference of control methods for any hazard. This order of preference for hazard control is repeated in various forms and names in safety publications ranging from machine safeguarding to process safety management. Regulations require the preferential use of engineering controls where practical to the use of PPE. Regulations requiring preferential engineering controls include: Occupational Noise Exposure (29 CFR 1910.95), [PPE] General Requirements (29 CFR 1910.132), Respiratory Protection (29 CFR 1910.134), and the standards for hazardous substances such as lead, asbestos, cadmium, et al., known as the vertical standards (29 CFR 1910, Subpart Z). The HRPS generally applicable to any system and type of hazard is provided in Figure 3.4. The HRPS in Figure 3.4 is ranked from most protective and most preferred at top to least protective and least preferred at bottom.

The HRPS should be used as a management tool to ensure hazards are adequately controlled. High-risk hazards should be eliminated or controlled using safety devices. Low-risk hazards are acceptably controlled by safe working procedures.

Determine if there is any way to perform the task with the circuits de-energized, or if some hazards could be eliminated through lockout. If the circuit can be locked out, the analysis is concluded. Make lockout a requirement for this task. If only part of the circuit can be locked out during conduct of this task or during a portion of this

Figure 3.4: Hazard Reduction Precedence Sequence (HRPS)

- Eliminate the hazard by design.

- Provide safety devices that act automatically to prevent the hazard occurence.

- Provide indications and warnings of an impending hazard so that operator action can be taken to prevent the hazard occurrence.

- Provide safe working procedures and training so that operator action can be taken to avoid hazard occurrence.

task, continue the analysis with the assumption that the circuit or part of the circuit will be locked out. Make lockout a requirement in the work procedure.

Once all hazards that can be eliminated through lockout have been identified, determine if insulated tools could be used to perform the task and whether they would provide adequate protection from shock. OSHA does not require the use of gloves when using insulated tools unless the hand is exposed to electric shock from sources other than the live part being worked on. In a letter of interpretation from OSHA to Mr. Fielding, dated May 20, 1996, OSHA stated,

> Insulated hand tools (which conform to International Electrotechnical Commission (IEC) 900 and American Society for Testing and Materials (ASTM) F 1505 standards) rated for the voltage involved would be considered insulation of the person from the energized part on which work is being performed. Generally, these tools which have a maximum rated voltage of 1000 volts for alternating current (a. c.) applications and 1500 volts for direct current (d. c.) applications would be suitable for work covered under the provisions of 29 CFR 1910 Subpart S.

In determining if insulating tools provide adequate protection, consider hazards to the head, hands, forearms, and feet. The bottom line is that if there is a credible hazard that cannot be eliminated or otherwise controlled, PPE capable of protecting the employee from the hazard must be identified and provided. If there is a chance that an employee could touch exposed live parts with his/her hands, insulating gloves rated for the voltage must be worn. When considering the opportunity to inadvertently touch exposed live parts, remember that accidents can be caused by loss of balance, being bumped by other people, or being distracted by loud noises. If live parts are exposed such that a finger or hand could contact them, then the hazard is credible. Where exposed live parts are recessed such that only a probe can contact them and not a body part, then the hazard is not credible.

When there is a risk of contacting exposed live parts with the forearm, then insulating sleeves must be worn. If an employee reaches past exposed live parts that can be contacted by the forearm, the hazard is credible. If the forearm cannot make contact with the exposed live parts, then the hazard is not credible.

When there is a risk of contacting exposed live parts with the head, insulating headgear must be worn. Remember that the insulating PPE must be rated for the voltage involved. Most hard hats do not have an insulating rating and so cannot be used for this purpose.

When there is a risk of step and touch potentials requiring PPE or when there is a risk of stepping on exposed live parts, dielectric overshoes are required.

If the insulating tool will provide adequate protection for a particular task, continue with a flash hazard analysis. If the insulating tool will not provide adequate protection by itself, continue the analysis with the assumption that insulated tools will be used where appropriate. Make the use of insulated tools a requirement of the work procedure.

Determine if it would be practical to wear insulating PPE. If not, the exposed live parts must be insulated. If PPE is practical, determine the type and rating needed based on the body part to be protected and the voltage.

Selection of Electrical PPE

The definitive guideline for electrical PPE can be found in the NFPA Standard for Electrical Safety Requirements for Employee Workplaces (NFPA 70E – 2000, Part II, Section 3-3). PPE may be needed for arc flash protection and shock protection. NFPA 70E provides more current information than OSHA regulations, especially concerning PPE for arc-flash protection.

PPE for Electrical Shock

Electrically-insulated PPE must be worn when there is a credible likelihood of electrical shock that is not otherwise controlled. In this context, a credible likelihood means that electrical shock is reasonably likely to occur. The mere presence of live electrical parts is not sufficient to establish a credible electric shock hazard for qualified employees. Qualified employees are trained and competent to work in the vicinity of live electrical parts. It is assumed that they are capable of taking adequate care not to intentionally touch live electrical parts with uninsulated tools. Live electrical parts that are recessed in insulated covers or molded parts such that they cannot be touched do not present a credible likelihood of electric shock for qualified employees. Likewise, live electrical parts covered with insulating blankets or temporary insulating covers do not continue to pose a hazard. The degree to which live electrical parts are guarded from inadvertent touch is a major consideration in determining the likelihood of electrical shock.

The work to be performed is also a major consideration for determining the likelihood of electric shock. The more involved the work, the more likely the qualified employee is to inadvertently touch live electrical parts. However, even if the qualified employee has no intention of getting close to live electrical parts, he can be bumped or startled into the live equipment. In high pedestrian or vehicle traffic areas, the chances of inadvertent collision and subsequent electrical shock are increased. Factors contributing to the likelihood of electrical shock due to being bumped or startled include the size of the electrical enclosure being worked on, the stability of the work platform, and the activity and noise levels in the area. If the qualified person stands on an unstable work platform, such as a portable ladder, the likelihood of electric shock increases.

Other considerations in determining the likelihood of electric shock include the posture of the qualified person and the level of illumination on the work. Where a qualified person must kneel, lay down, or reach awkwardly, the likelihood of inadvertently touching live parts increases. Where the qualified person cannot see his work or the surrounding parts clearly, the likelihood of electric shock increases. All of these considerations combine to affect the likelihood of electric shock.

The qualified employee and/or the supervisor must consider these factors to determine if PPE for electric shock is warranted and what PPE, if any, is needed to protect the worker. This is a subjective evaluation of likelihood. It can be formalized with a weighted checklist, but need not be depending on the confidence management has in the sound judgement of the decision-makers. An example of one such weighted checklist is given in Figure 3.5.

Electrically-insulated gloves are rated for certain maximum voltages. Class 00 gloves are rated for a maximum of 500 V. Class 0 gloves are rated for a maximum of 1000 V. The appropriate gloves must be worn where protection from an electrical shock hazard to the hand and wrist is needed. Long-cuff gloves are available to protect the lower forearm. Insulating sleeves are required if the upper forearm needs protection from electric shock because of reaching past exposed live parts. Covering exposed live parts with insulating blankets or temporary insulating covers provides equivalent protection to wearing PPE and affords the qualified employee maximum manual dexterity and comfort.

Head protection must be worn where there is a need to protect the qualified employee from electrical shock hazards at head height. Such head protection must have an appropriate insulation rating. A hard hat or bump cap may not have an insulation rating provided by the manufacturer because it is designed to provide protection from impact hazards, not from electric shock. Electrically insulated head protection is available in the form of insulating hoods and as helmets. Class G helmets are suitable for low voltage protection (600 V or less). Class E helmets are suitable for high voltage protection (> 600 V). The voltages at which these helmets are tested are much greater than the voltages they protect against.

Are live electrical parts exposed to inadvertent touch or contact with uninsulated tools?

◆　　If not, insulated PPE is not needed.

◆　　If yes, answer the following questions. Where the factors add up to 10 or more, protective insulation or insulated PPE is required.

1.　Does the task require reaching past exposed live parts within their restricted boundary? (10)
2.　Is the task more complicated than simple voltage measurements using an appropriately insulated probe? (10)
3.　Are there exposed live parts that can be contacted with the head? (10)
4.　Is there a possibility that the qualified employee could be bumped into exposed live parts? (10)
5.　Is the qualified employee standing on an unstable work platform, ladder, or slippery surface? (10)
6.　Is there a possibility that the qualified employee could be startled by loud impulse noises that are known to occur in the area? (5)
7.　Is the qualified employee required to kneel, squat, or lay down to perform the task? (5)
8.　Is the qualified employee required to work with the hands above shoulder height for more than 4 minutes? (5)
9.　Is the qualified employee required to use excessive force? (5)
10.　Is the qualified employee required to work with the hands above eye height? (5)
11.　Is the work space inadequately illuminated? (5)
12.　Does the work require simultaneous use of both hands? (2)

Figure 3.5: Insulated PPE Checklist

PPE for Arc Flash

Essentially, safety glasses and clothing that does not melt or readily ignite are required to protect employees from arc flash. The characteristics of clothing and the extent to which it must be fire retardant is dependent on the voltage and the available short circuit current.

Available short circuit current is the current that could result from bolting all three phases together at a given point in the electrical distribution system in the facility. This type of short circuit is used in the calculations because it generally results in the maximum short circuit current and because it simplifies the calculations. The available short circuit current is important to calculate because it determines the choice of electrical installation components, overcurrent protection devices, and PPE. Electrical system components must be selected to withstand the mechanical and thermal stresses caused by a short circuit. Overcurrent protection devices must be rated to withstand the maximum short circuit current and to be sensitive enough to detect the minimum expected short circuit current. When working on or near exposed live parts, there is always a risk of causing a short circuit. When a short circuit results from bringing an object in contact with energized parts, an arc flash will occur. The amount of energy released by the arc flash is directly dependent on the short circuit current. Therefore, it is essential to choose PPE for this potential hazard based on the calculated available short circuit current.

The available short circuit current is typically calculated by dividing the nominal voltage at the fault location by the effective circuit impedance. Calculation of the circuit impedance is beyond the scope of this book. Guidance for calculating the available short circuit current is given in the Institute of Electrical and Electronics Engineers (IEEE) publication, *Recommended Practice for Electrical Power Distribution for Industrial Plants* (IEEE Std 141 - 1993), Chapter 4. However, the safety professional should have a general understanding of scope of the task.

Many factors contribute to the available short circuit current at any given point. Because of their large current capacity, the utility generators that supply power to the facility are generally unaffected by short circuits in the facility. The electric utility company will provide the impedance value at the service point of the facility. Generators located at the facility likewise contribute to the available short circuit current. Less intuitive is the fact that motors also contribute to the available short circuit current. When power is removed from a synchronous motor, the momentum of the motor keeps it turning. As it slows down, the motor acts like a generator. The manufacturers of generators and motors can provide their impedance values. The busbars, cables, and other conductors also contribute to the impedance in the form of resistance. The manufacturers of conductors can provide their impedance values. The calculation then takes these component impedance values and combines them depending on whether they are in series or parallel with each other relative to the fault location. The available short circuit current will not be the same at all locations in the electrical system.

For the purposes of PPE determination, calculations of available short circuit current need to be made at various points in the electrical system. If work on or near exposed live parts will not be permitted, then calculations are not required at that point for the purposes of determining appropriate PPE. Calculate the short circuit current at the load with the highest impedance. Where multiple 3-phase voltages are distributed, calculate the short circuit current available at the loads with the highest impedance for each nominal voltage. Once the available short circuit current is calculated, the appropriate PPE for protection against arc flash can be chosen.

NFPA 70E, Part II, Table 3-3.9.1 (see Table 3.1) assigns a hazard category to certain routine tasks involving electrical equipment of various voltages. NFPA 70E, Part II, Table 3-3.9.2 (see Table 3.2) identifies the PPE for the various hazard categories. NFPA 70E, Part II, Table 3-3.9.3 (see Table 3.3) describes the required flame resistance (FR) of various clothing in terms of their Arc Thermal Performance Exposure Value (ATPV). These tables have been provided to help explain the process of determining appropriate PPE and to introduce the NFPA 70E standard. It is strongly recommended that the standard be consulted with all of its detailed explanation and requirements.

Table 3-3.9.1 lists various tasks on or near exposed live parts at various voltages. To determine what PPE is required, find the type of equipment on the table that most closely matches the subject equipment. Note 3 and Note 6 of the table allow the arc flash hazard to be reduced one level if the available short circuit current is less than a specified amount. One of these notes is applicable to every low voltage task (600 V or less). Determine what level of arc flash hazard is presented by the task. The need for voltage-rated gloves and tools are also indicated on the table next to the arc flash hazard category. Using Table 3-3.9.2, determine what PPE is required for the arc flash hazard level identified. Using Table 3-3.9.3, determine what the ATPV is required to be for certain clothing to be considered adequate for the hazard level.

PPE for arc flash on equipment rated for 600 V or less is largely dependent on the type of equipment, its short circuit current rating, and the available short circuit current. Types of equipment listed include panelboards, switchboards, motor control centers (MCC), and switchgear. A panelboard is an electrical cabinet or box containing conductors, protection devices, and possibly switches. Panelboards are designed to be opened only from the front. An example of a panelboard is a circuit breaker panel. A switchboard is an electrical cabinet or box containing conductors, protection devices, control switches, and instrumentation. A switchboard is usually designed to be opened from more than one side. An example of a switchboard is a programmable logic controller cabinet. A MCC is self explanatory; it is a cabinet containing the motor control circuitry. Switchgear means an electrical cabinet containing the switching and disconnecting means for the utility service, power station, or substation.

Most tasks at 480 V are considered hazard level 0 or 1. Hazard level 1 requires a FR long sleeve shirt and pants or FR coveralls, a hardhat, and safety glasses. Regular weight (minimum 12 oz/yd^2 fabric weight), untreated, denim cotton blue jeans are acceptable in lieu of FR pants. FR material for hazard level 1 must have an ATPV of at least 5.

Working on energized parts inside panelboards or switchboards rated from 240 V to 600 V is considered hazard level 2, unless the available short circuit current is less than 10 kA. This includes voltage testing. Hazard level 2 requires FR coveralls or pants and shirt with an ATPV of 5 over cotton T-shirt and regular pants or FR coveralls or pants and shirt with an ATPV of 8. Additionally, a hardhat, safety glasses or goggles, leather glove protectors, leather shoes, hearing protection, and a double-layer switching hood are required for the tasks mentioned.

Insertion or removal of individual starter "buckets" from MCC, insertion or removal (racking) of circuit breakers from cubicles with doors open on 600 V class switchgear, and removal of bolted covers to expose bare, energized parts on 600 V class switchgear are considered hazard level 3. However, if the available short circuit current is less than 10 kA for the MCC or less than 25 kA for the 600 V class switchgear, the hazard level can be reduced to 2. Hazard level 3 requires three layers of clothing. Two FR coveralls (with an ATPV of 5) over cotton T-shirt and regular pants, or FR coverall over FR pants over cotton T-shirt and regular pants. Additionally, a hardhat with FR liner, safety glasses or goggles, leather glove protectors, leather shoes, hearing protection, and a double-layer switching hood are required.

Requirements for Inspection of Electrical PPE

Electrical PPE must be inspected frequently to ensure its safety. Qualified employees should inspect their electrical PPE before each use and after any incident that might have damaged it. The purpose of such an inspection is to ensure the PPE remains an intact, complete barrier to current. Inspections should look for:

- A hole, tear, puncture, or cut
- Ozone cutting or ozone checking (the cutting action produced by ozone on rubber under mechanical stress into a series of interlacing cracks)
- An embedded foreign object
- Any of the following texture changes: swelling, softening, hardening, or becoming sticky or inelastic
- Any other defect that damages the insulating properties

The qualified employee must also conduct an air test of electrically insulating gloves along with each inspection. An air test is made by sealing the glove by holding the wrist opening shut and squeezing the glove to see if it will hold air.

In order to preserve electrical PPE and insulating materials and tools, employees must be trained in their proper care. Insulating PPE and barriers must be protected against cuts and punctures. Insulating tools must be protected from cracking. To that end, employees must be careful how the equipment is handled when in use. Protector gloves must be worn over insulating gloves, except as follows [taken from 29 CFR 1910.137]:

1. Protector gloves need not be used with Class 0 gloves, under limited-use conditions, where small equipment and parts manipulation necessitate unusually high finger dexterity.

2. Any other class of glove may be used for similar work without protector gloves if the employer can demonstrate that the possibility of physical damage to the gloves is small and if the class of glove is one class higher than that required for the voltage involved.

3. Extra care is needed in the visual examination of the glove and in the avoidance of handling sharp objects.

4. Insulating gloves that have been used without protector gloves may not be used at a higher voltage until they have been tested.

Insulating equipment must be stored only in locations free from sunlight or UV radiation, extreme heat or cold, humidity, ozone, solvent vapors, or other chemicals that are incompatible with the insulating material. Insulating equipment must be kept clean of foreign materials.

Requirements for Testing Electrical PPE and Insulating Equipment

Electrical insulating gloves must be tested using standard test methods to ensure their insulating capability. Gloves issued for use must be tested every 6 months. Gloves not issued for use (i.e., in the stock room) must be tested every 12 months. Some glove manufacturers put the date of manufacture or expiration date on the gloves. Other manufacturers put a manufacturing code on the gloves by which they can tell when they were manufactured.

Unless a large number of gloves are continuously in use, it may be more economical to dispose of gloves that have expired than to test them. For economy to be maximized, stock only as many gloves as are needed, and issue gloves only as needed. To ensure compliance with testing requirements keep inventory records of the expiration dates of gloves so that stocks and issued gloves can be collected for disposal or testing.

Other insulating gear, such as mats, blankets, and insulating tools, must be tested annually.

Table 3.1: NFPA 70E, Part II, Table 3-3.9.1 Hazard Risk Category Classifications

Task (Assumes Equipment Is Energized, and Work Is Done Within the Flash Protection Boundary)	Hazard/ Risk Category	V-rated Gloves	V-rated Tools
Panelboards rated 240 V and below – Notes 1 and 3	—	—	—
Circuit breaker (CB) or fused switch operation with covers on	0	N	N
CB or fused switch operation with covers off	0	N	N
Work on energized parts, including voltage testing	1	Y	Y
Remove/install CBs or fused switches	1	Y	Y
Removal of bolted covers (to expose bare, energized parts)	1	N	N
Opening hinged covers (to expose bare, energized parts)	0	N	N
Panelboards or Switchboards rated >240 V and up to 600 V (with molded case or insulated case circuit breakers) — Notes 1 and 3	—	—	—
CB or fused switch operation with covers on	0	N	N
CB or fused switch operation with covers off	1	N	N
Work on energized parts, including voltage testing	2*	Y	Y
600 V Class Motor Control Centers (MCCs) – Notes 2 (except as indicated) and 3	—	—	—
CB or fused switch or starter operation with enclosure doors closed	0	N	N
Reading a panel meter while operating a meter switch	0	N	N
CB or fused switch or starter operation with enclosure doors open	1	N	N
Work on energized parts, including voltage testing	2*	Y	Y
Work on control circuits with energized parts 120 V or below, exposed	0	Y	Y
Work on control circuits with energized parts >120 V exposed	2*	Y	Y
Insertion or removal of individual starter "buckets" from MCC — Note 4	3	Y	N
Application of safety grounds, after voltage test	2*	Y	N
Removal of bolted covers (to expose bare, energized parts)	2*	N	N
Opening hinged covers (to expose bare, energized parts)	1	N	N
600 V Class Switchgear (with power circuit breakers or fused switches) — Notes 5 and 6	—	—	—
CB or fused switch operation with enclosure doors closed	0	N	N
Reading a panel meter while operating a meter switch	0	N	N
CB or fused switch operation with enclosure doors open	1	N	N
Work on energized parts, including voltage testing	2*	Y	Y
Work on control circuits with energized parts 120 V or below, exposed	0	Y	Y
Work on control circuits with energized parts >120 V exposed	2*	Y	Y
Insertion or removal (racking) of CBs from cubicles, doors open	3	N	N
Insertion or removal (racking) of CBs from cubicles, doors closed	2	N	N
Application of safety grounds, after voltage test	2*	Y	N
Removal of bolted covers (to expose bare, energized parts)	3	N	N
Opening hinged covers (to expose bare, energized parts)	2	N	N
Other 600 V Class (277 V through 600 V, nominal) Equipment — Note 3	—	—	—
Lighting or small power transformers (600 V, maximum)	—	—	—
Removal of bolted covers (to expose bare, energized parts)	2*	N	N
Opening hinged covers (to expose bare, energized parts)	1	N	N
Work on energized parts, including voltage testing	2*	Y	Y
Application of safety grounds, after voltage test	2*	Y	N

Task (Assumes Equipment Is Energized, and Work Is Done Within the Flash Protection Boundary)	Hazard/ Risk Category	V-rated Gloves	V-rated Tools
Revenue meters (kW-hour, at primary voltage and current)	—	—	—
Insertion or removal	2*	Y	N
Cable trough or tray cover removal or installation	1	N	N
Miscellaneous equipment cover removal or installation	1	N	N
Work on energized parts, including voltage testing	2*	Y	Y
Application of safety grounds, after voltage test	2*	Y	N
NEMA E2 (fused contactor) Motor Starters, 2.3 kV through 7.2 kV	—	—	—
Contactor operation with enclosure doors closed	0	N	N
Reading a panel meter while operating a meter switch	0	N	N
Contactor operation with enclosure doors open	2*	N	N
Work on energized parts, including voltage testing	3	Y	Y
Work on control circuits with energized parts 120 V or below, exposed	0	Y	Y
Work on control circuits with energized parts >120 V, exposed	3	Y	Y
Insertion or removal (racking) of starters from cubicles, doors open	3	N	N
Insertion or removal (racking) of starters from cubicles, doors closed	2	N	N
Application of safety grounds, after voltage test	3	Y	N
Removal of bolted covers (to expose bare, energized parts)	4	N	N
Opening hinged covers (to expose bare, energized parts)	3	N	N
Metal Clad Switchgear, 1 kV and above	—	—	—
CB or fused switch operation with enclosure doors closed	2	N	N
Reading a panel meter while operating a meter switch	0	N	N
CB or fused switch operation with enclosure doors open	4	N	N
Work on energized parts, including voltage testing	4	Y	Y
Work on control circuits with energized parts 120 V or below, exposed	2	Y	Y
Work on control circuits with energized parts >120 V, exposed	4	Y	Y
Insertion or removal (racking) of CBs from cubicles, doors open	4	N	N
Insertion or removal (racking) of CBs from cubicles, doors closed	2	N	N
Application of safety grounds, after voltage test	4	Y	N
Removal of bolted covers (to expose bare, energized parts)	4	N	N
Opening hinged covers (to expose bare, energized parts)	3	N	N
Opening voltage transformer or control power transformer compartments	4	N	N
Other Equipment 1 kV and above	—	—	—
Metal clad load interrupter switches, fused or unfused	—	—	—
Switch operation, doors closed	2	N	N
Work on energized parts, including voltage testing	4	Y	Y
Removal of bolted covers (to expose bare, energized parts)	4	N	N
Opening hinged covers (to expose bare, energized parts)	3	N	N
Outdoor disconnect switch operation (hookstick operated)	3	Y	Y
Outdoor disconnect switch operation (gang-operated, from grade)	2	N	N
Insulated cable examination, in manhole or other confined space	4	Y	N
Insulated cable examination, in open area	2	Y	N

Legend:
V-rated Gloves are gloves rated and tested for the maximum line-to-line voltage upon which work will be done.
V-rated Tools are tools rated and tested for the maximum line-to-line voltage upon which work will be done.
2* means that a double-layer switching hood and hearing protection are required for this task in addition to the other Hazard/Risk Category 2 requirements of Table 3-3.9.2 of Part II.
Y = yes (required)
N = no (not required)

Notes:
1. 25 kA short circuit current available, 0.03 second (2 cycle) fault clearing time.
2. 65 kA short circuit current available, 0.03 second (2 cycle) fault clearing time.
3. For < 10 kA short circuit current available, the Hazard/Risk Category required may be reduced by one number.
4. 65 kA short circuit current available, 0.33 second (20 cycle) fault clearing time.
5. 65 kA short circuit current available, up to 1.0 second (60 cycle) fault clearing time.
6. For < 25 kA short circuit current available, the Hazard/Risk Category required may be reduced by one number.

Table 3.2: NFPA 70E, Part II, Table 3-3.9.2 Protective Clothing and Personal Protective Equipment (PPE) Matrix

Protective Clothing & Equipment	Hazard/Risk Category Number					
	-1 (Note 3)	0	1	2	3	4
Untreated Natural Fiber	—	—	—	—	—	—
a. T-shirt (short-sleeve)	X			X	X	X
b. Shirt (long-sleeve)	X					
c. Pants (long)	X	X	X (Note 4)	X (Note 6)	X	X
FR Clothing (Note 1)	—	—	—	—	—	—
a. Long-sleeve shirt			X	X	X (Note 9)	X
b. Pants			X (Note 4)	X (Note 6)	X (Note 9)	X
c. Coverall			(Note 5)	(Note 7)	X (Note 9)	(Note 5)
d. Jacket, parka, or rainwear			AN	AN	AN	AN
FR Protective Equipment	—	—	—	—	—	—
a. Flash suit jacket (2-layer)						X
b. Flash suit pants (2-layer)						X
Head protection	—	—	—	—	—	—
a. Hard hat			X	X	X	X
b. FR hard hat liner					X	X
Eye protection	—	—	—	—	—	—
a. Safety glasses	X	X	X	AL	AL	AL
b. Safety goggles				AL	AL	AL
Face protection double-layer switching hood				AR (Note 8)	X	X
Hearing protection (ear canal inserts)				AR (Note 8)	X	X
Leather gloves (Note 2)			AN	X	X	X
Leather work shoes			AN	X	X	X

Legend:
AN = As needed
AL = Select one in group
AR = As required
X = Minimum required

Notes:
1. See Table 3-3.9.3. (ATPV is the Arc Thermal Performance Exposure Value for a garment in cal/cm^2.)
2. If voltage-rated gloves are required, the leather protectors worn external to the rubber gloves satisfy this requirement.
3. Class -1 is only defined if determined by Notes 3 or 6 of Table 3-3.9.1 of Part II.
4. Regular weight (minimum 12 oz/yd^2 fabric weight), untreated, denim cotton blue jeans are acceptable in lieu of FR pants. The FR pants used for Hazard/Risk Category 1 shall have a minimum ATPV of 5.
5. Alternate is to use FR coveralls (minimum ATPV of 5) instead of FR shirt and FR pants.

6. If the FR pants have a minimum ATPV of 8, long pants of untreated natural fiber are not required beneath the FR pants.

7. Alternate is to use FR coveralls (minimum ATPV of 5) over untreated natural fiber pants and T-shirt.

8. A double-layer switching hood and hearing protection are required for the tasks designated 2* in Table 3-3.9.1 of Part II.

9. Alternate is to use two sets of FR coveralls (each with a minimum ATPV of 5) over untreated natural fiber clothing, instead of FR coveralls over FR shirt and FR pants over untreated natural fiber clothing.

Table 3.3: NFPA 70E, Part II, Table 3-3.9.3 Protective Clothing Characteristics

Typical Protective Clothing Systems

Hazard Risk Category	Clothing Description (Number of clothing layers is given in parentheses)	Total Weight oz/yd²	Minimum Arc Thermal Performance Exposure Value (ATPV)* or Breakopen Threshold Energy (E_{BT})* Rating of PPE cal/cm²
0	Untreated cotton (1)	4.5 – 7	N/A
1	FR shirt and FR pants (1)	4.5 – 8	5
2	Cotton underwear plus FR shirt and FR pants (2)	9 – 12	8
3	Cotton underwear plus FR shirt and FR pants plus FR coverall (3)	16 – 20	25
4	Cotton underwear plus FR shirt and FR pants plus double layer switching coat and pants (4)	24 – 30	40

*ATPV is defined in the ASTM P S58 standard arc test method for flame resistant (FR) fabrics as the incident energy that would just cause the onset of a second degree burn (1.2 cal/cm²). E_{BT} is reported according to ASTM P S58 and is defined as the highest incident energy which did not cause FR fabric breakopen and did not exceed the second-degree burn criteria. E_{BT} is reported when ATPV cannot be measured due to FR fabric breakopen.

Chapter Four

Developing Safe Work Procedures

Overview

Safe work procedures for electrical tasks start with safety policies. Procedures for working on de-energized electrical equipment are developed through the Control of Hazardous Energy or Lockout/ Tagout program. Three regulations apply to general industry working on de-energized electrical equipment: the Control of Hazardous Energy (Lockout/Tagout) standard (29 CFR 1910.147), the standard for Selection and Use of [Electrical] Work Practices (29 CFR 1910.333), and Electric Power Generation, Transmission, and Distribution (29 CFR 1910.269(d)).

It is essential that there be adequate planning and procedures in place prior to commencement of work on or near exposed live parts. Careful planning and procedures, properly carried out, can greatly reduce the risk associated with such work. Planning includes determining how close to exposed live parts or conductors the work will be performed. The proximity to overhead lines and fixed exposed live parts must be identified. The flash protection boundary, limited approach boundary, restricted approach boundary, and prohibited approach boundary should be determined. Appropriate safety measures to control the hazards of electric shock and arc flash should be identified in the work plan.

De-energized Electrical Equipment

The Lockout/Tagout Standard

The Lockout/Tagout standard is not applicable to certain types of tasks—namely, certain hot tap operations, certain maintenance and servicing integral to production, and minor tool changes. However, it falls to the employer to provide safe work procedures, training, tools, and PPE as appropriate to safeguard employees from hazards while performing these tasks.

Within the applicability of the Lockout/Tagout standard, machine-specific energy control procedures must be developed for all equipment unless it meets certain conditions. Cord and plug equipment is exempted from the standard. Equipment that meets the following criteria does not require a written lockout/tagout procedure:

- The machine or equipment has no potential for stored energy, residual energy, or the reaccumulation of stored energy after shut down which could endanger employees.

- The machine or equipment has a single energy source which can be readily identified and isolated.

- The isolation and locking out of that energy source will completely de-energize and deactivate the machine or equipment.

Figure 4.1: Lockout Device

- The machine or equipment is isolated from that energy source and locked out during servicing or maintenance.

A lockout device is a lock or device facilitating the hanging of a lock on an energy isolation device that is not otherwise lockable (see Figure 4.1).

- A single lockout device will achieve a locked-out condition.

- The lockout device is under the exclusive control of the authorized employee performing the servicing or maintenance.

An energy isolation device is a mechanism which prevents energy from reaching the machine (see Figure 4.2). An electrical disconnect, circuit breaker, fuse, or isolation valve are examples of energy isolation devices.

- The servicing or maintenance does not create hazards for other employees.

- The employer, in utilizing this exception, has had no accidents involving the unexpected activation or re-energization of the machine or equipment during servicing or maintenance.

Each required lockout/tagout procedure must have the following elements:

- A specific statement of the intended use of the procedure.

- Specific procedural steps for shutting down, isolating, blocking, and otherwise securing machines or equipment to control hazardous energy.

Figure 4.2: Energy Isolation Device

- Specific procedural steps for the placement, removal, and transfer of lockout/tagout devices and the responsibility for them.
- Specific requirements for testing a machine or equipment to determine and verify the effectiveness of lockout/tagout devices and other energy control measures.

The machine-specific energy isolation procedures should be maintained at the applicable machines where they are readily accessible to authorized employees. Authorized employees performing lockout/tagout must follow the applicable procedure.

When changes to equipment or machinery affect energy isolation, the machine-specific energy isolation procedures must reflect those changes. All authorized employees subject to use a procedure and all the procedure's affected employees must understand the changes.

Tags are not acceptable alternatives to locks unless the energy isolation device is not capable of being locked out. If hanging a tag without a lock, the employer must consider additional safety measures to reduce the likelihood of inadvertent energization. For example, the removal of an isolating circuit element, the blocking of a controlling switch, the opening of an extra disconnecting device, or the removal of a valve handle.

Do not use control circuit devices (e.g., push buttons, selector switches, and interlocks) as the sole means for de-energizing circuits or equipment. Do not use interlocks for electrical equipment as a substitute for lockout and tagout procedures.

Seven Steps for Lockout/Tagout

The Lockout/Tagout standard requires that these seven steps be followed when performing any lockout. These steps should be performed in conjunction with machine-specific energy isolation procedures.

1. Identify all energy sources, their hazards, and the means for isolating or relieving each energy source.

2. Notify all affected employees that the equipment or machinery will be shut down and energy sources locked out.

3. Shut down the machine or equipment per standard operating procedures.

4. Isolate each energy source using the required energy isolation device in its required "safe" position in accordance with machine-specific energy isolation procedures.

   Mechanical disconnect may be necessary to adequately isolate the energy sources.

5. Place a lockout lock with or without an additional lockout device as necessary on each energy isolation device. Completion of a tag in addition to each lock is recommended by the Lockout/ Tagout standard and required by the Selection and Use of [Electrical] Work Practices standard. If the energy isolation device is not capable of being locked out, secure a completed tag to the energy isolation device.

Using a lockout lock without a tag for electrical maintenance is acceptable only when <u>all</u> of the following conditions are met:

• Only one circuit or piece of equipment is de-energized.

Note that each authorized employee working on the machine must place an individual lock or tag at each energy isolation device (unless a group lockout system is in use).

• The lockout period does not extend beyond the work shift.

• Employees exposed to the hazards associated with re-energizing the circuit or equipment are familiar with this procedure.

   Where an energy isolation device is not capable of being locked out, the employer must modify it or replace it such that it will accept a lockout device whenever the machine or subject equipment is replaced or undergoes major repair, renovation, or modification. Newly installed machines and equipment must be capable of being locked out.

6. Relieve any residual or stored energy, including electrical energy, that may be stored in capacitors or large inductors. Hydraulic and pneumatic pressure which remains in the system piping when a three-position valve returns to neutral is potentially dangerous stored energy. Place rams and moving parts in their lowest energy state or block them in position if they pose a hazard to personnel.

7. Verify that energy sources have been correctly isolated by performing <u>all</u> of the following, as applicable:

• Visually check that all energy isolation devices are in their required "safe" positions in accordance with the machine-specific energy isolation procedure (valves are in their required "safe" position, required blanks are in place between pipe flanges in pipelines, and pipes required to be disconnected are misaligned or removed).

• Attempt to start the equipment at an operating switch or control panel. An additional person may be needed to verify that the equipment is not energized from the control panel when the control panel is remote from the equipment. Be sure to return the switch to the OFF or neutral position after verification.

• Ensure that a person qualified to work on energized equipment tests the circuit elements and electrical parts of equipment to which employees will be exposed and verifies that the circuit elements and equipment parts are de-energized. The test must also determine if any energized condition exists as a result of inadvertently induced voltage or unrelated voltage feedback, even though specific parts of the circuit have been de-energized and presumed to be safe. If the circuit to be tested is over 600 volts, nominal, the test equipment must be checked for proper operation immediately before and immediately after this test. This is especially important when a motor has tripped out or failed.

Release from Lockout/Tagout

Follow this procedure to restore energy that is locked and tagged out. Use these steps in conjunction with machine-specific energy isolation procedures.

1. Ensure that the machine or equipment is ready to be restored by:

  • Removing all tools, parts, and other material from the work area and machine.

  • Replacing all covers, guards, panels, and other enclosures.

2. Notify all affected employees that locks are going to be removed, and ensure that all personnel are clear of the machine and at a safe distance or in a safe location.

3. Remove all lockout devices and tags, and restore energy isolation devices to their required pre-startup positions in accordance to the machine-specific energy isolation procedure.

4. After lockout devices and tags have been removed, notify all affected employees that the machine or equipment is no longer locked and tagged out.

Additional Lockout/Tagout Procedures

The lockout/tagout program must include procedures for group lockout, shift change, removal of another employee's lockout/tagout device, and how to coordinate lockout with contractors.

Periodic Inspections

As needed and prudent, at least annually, the employer will direct a periodic inspection. The periodic inspection must be conducted by an authorized employee other than the one being observed. Each machine-specific procedure must be inspected at least annually. Where tags are used without a lock, more frequent periodic inspections are required.

The inspector will identify and document all deviations from, or inadequacies in, the procedures for subsequent correction. The inspector will review the responsibilities of authorized employees with each employee who is subject to use the procedure being inspected. Where tags are used, the inspector will additionally review with each affected employee his/her understanding of the following:

  • Tags are warning devices, and do not provide the physical protection that is provided by a lockout lock with or without additional lockout devices.

  • Tags may give a false sense of security.

  • Tags must be highly visible, legible, and understandable to be effective.

  • Tags and their means of attachment must withstand the environmental conditions in the workplace.

  • The means of attaching tags must prevent accidental detachment.

  • Employees may not remove tags belonging to another employee without proper permission and following appropriate procedures. Employees may never bypass, ignore, or otherwise defeat a tag.

OSHA has authored letters of interpretation and abatement agreements that allow inspection of a representative number of authorized employees, instead of all employees subject to use the procedure. Therefore, for each machine-specific procedure, a representative number of authorized employees that use the procedure must be inspected. A scheme for determining what constitutes a representative number of employees is shown in Table 4.1.

Table 4.1: Suggested Minimum Sample Size for Population

Size of Population	Extremely Important to Verify Compliance	Additional Information to Substantiate Compliance	Ancillary Information to Verify Overall Compliance
2-10	100%	100%	30%
11-25	100%	40%	20%
26-50	50%	20%	15%
51-100	25%	10%	8%
101-250	15%	7%	5%
251-500	10%	5%	3%
501-1000	5%	3%	2%
Over 1000	3%	2%	1%

For a facility with hundreds of machine-specific procedures and a handful of authorized employees, the task of conducting periodic inspections could be extremely difficult. It is essential that machine-specific procedures be written to apply as broadly as possible while maintaining the required accuracy and completeness. Furthermore, a tracking system of all procedures and how many employees are authorized to use the procedure should be developed at the beginning of the year. Throughout the year, when a procedure is first used, it should be inspected and documented. After six months, most of the required inspections will have been completed. Do not wait till the last quarter of the year to try to complete all the required periodic inspections.

The tracking of completed periodic inspections is important because OSHA requires that the employer certify all required inspections have been done. The certification must identify the machine or equipment on which the energy isolation procedure was being utilized, the date of the inspection, the employees included in the inspection, and the person performing the inspection.

Lockout/Tagout Devices

Locks must be sturdy, standardized within the facility by color, shape, or size, and identifiable as being hung by a specific authorized employee. Locks must be in the exclusive control of the person who hung them. Therefore, extra keys should be destroyed. If extra keys or master keys are kept, a key control program must be implemented to provide exclusivity of control. This type of program is so hard to implement properly that it is not recommended. The cost of cutting off a lock in the event that an authorized employee is not available to remove it is cheap compared to OSHA determining that the lockout program does not provide required exclusivity of control.

Tags must also be sturdy, and standardized within the facility by color, shape, or size, print, and format. It must take 50 pounds or more force to remove a tag. Tags must identify the employee hanging the tag, warn against hazardous conditions if the machine or equipment is energized, and include key wording such as: DO NOT START, DO NOT OPEN, DO NOT CLOSE, DO NOT ENERGIZE, or DO NOT OPERATE. See Figure 4.3 for sample tags.

Since lockout devices are much safer than tags, a variety of lockout devices have been developed commercially to lock out many energy isolation devices that otherwise would not be capable of being locked out by a keyed padlock. These lockout devices can be used on many types of energy isolation devices or designed for a specific size and shape of energy isolation device. Lockout devices have been developed and are available commercially to lock out electrical cords and plugs (see Figure 4.4), circuit breakers (see Figure 4.5), fuse holders, gate valves, ball valves, butterfly valves, and wall switches. Other lockout devices have been developed to permit multiple lockouts or group lockout (e.g., hasps and lock boxes). Energy isolation devices themselves have been developed to incorporate locking features such as locking pneumatic in-line valves, locking pneumatic isolation valves that automatically depressurize lines downstream, and locking emergency stop buttons.

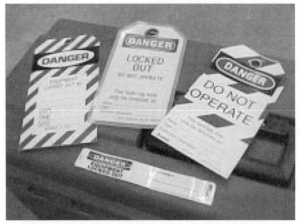

Figure 4.3: Examples of Tagout Devices

Figure 4.4: Cord and Plug Lockout Devices

Figure 4.5: Circuit Breaker Lockout Devices

Training for Maintenance Personnel

The lockout/tagout program has its own training requirements. Each employee working on a piece of equipment where unexpected startup or energization could cause injury must place his/her personal locks and tags on the equipment (or use a group lockout procedure). All maintenance personnel will probably be required to perform lockout. As such they will be authorized personnel under the standard. Authorized employees must be trained to know:

•   The types and magnitude of hazardous energy they may encounter in the facility.

•   The specific hazardous energy sources they will encounter when working on a specific machine or equipment.

- How to isolate energy per the machine-specific procedures for machines and equipment for which they are responsible.

- The necessity for reviewing and using machine-specific energy isolation procedures before work begins.

- How to complete and when to use tags.

- The procedure to remove lockout/tagout devices and re-energize machines for adjustment and testing.

- The procedure and prerequisites for removing another employee's lockout/tagout device.

- How to perform the Shift Change procedure (if applicable).

Additionally, authorized employees should be periodically reminded of the following information:

- Tags may give a false sense of security; tags do not prevent energy from being unintentionally or intentionally released.

- Tags must be highly visible, legible, and understandable in order to be effective.

- To remain effective, tags and their means of attachment must be constructed of materials sturdy enough to withstand the environment (e.g., laminated or plastic tags with aluminum, stainless steel, or plastic ties). Unprotected paper card stock with cloth string is unacceptable. Metal ties adjacent to electrical installations are hazardous.

Incorporating Electrical Safety Requirements into Lockout/Tagout Programs

It is more efficient to have a single program for working on de-energized equipment than one program for mechanical maintenance and a separate program for electrical maintenance. However, Subpart S differs from the Lockout/Tagout standard on a few points. When a single program for lockout is used, these differences must be incorporated.

- The electrical safety standard requires that a lock and a tag be placed on all energy isolation devices needed to protect personnel performing electrical maintenance or servicing. A lock alone will not suffice.

- A tag may be placed without a lock, but a second means of protection is required, such as tagging out a second energy isolation device upstream or disconnecting the conductors downstream of the energy isolation device.

- To verify the electrical equipment is de-energized, the circuit must be tested with an electrometer in addition to attempting to start the equipment. If testing circuits over 600 V, the electrometer must be immediately tested on a live circuit of known voltage after verifying that the subject circuit is de-energized. This test on a live circuit gives positive indication that the meter is functioning properly.

If electric power generation equipment is present in the facility, compliance with 29 CFR 1910.147 for such equipment is considered compliance with 29 CFR 1910.269(d). If transmission lines, substations, or switching stations (i.e., distribution equipment) will be worked on by employees, compliance with 29 CFR 1910.269(m) is required. As long as the requirements above are incorporated into the Lockout/Tagout program, a single program can be sufficient to comply with the standards.

When ascertaining if all required procedures have been developed, consider electrical maintenance inside electrical cabinets and on premises wiring. The electrical disconnect used to isolate and lockout equipment is often located on the equipment's electrical control panel. However, the supply side of the electrical disconnect is not isolated by placing it in the OFF position. Thus, placing the electrical disconnect in the OFF position does not protect employees from exposed live parts when working inside the cabinet. The lockout/tagout procedure should specify what equipment it isolates or does not isolate. For example, the lockout/tagout procedure for a CNC machine should state that it is not intended for work inside the main control panel for the machine. To lockout the main control panel, the employee must lockout the circuit breaker(s), feeder disconnect(s), or fuses that supply the main control panel. If the energy isolation point for the main control panel is not readily identifiable, a procedure is required by the Lockout/Tagout standard. Likewise, when a control cabinet is fed by more than one source of electrical power, de-energizing the control cabinet requires a procedure to identify all electrical sources and their isolation points.

Electrical servicing and maintenance or installation of new equipment is often performed on premises wiring not associated with a particular piece of equipment. Such equipment as circuit breaker panels, feeders, and branch circuits typically would require a written procedure because they could not meet the 8 criteria listed above.

Because of the non-routine nature of servicing and maintenance on premises wiring, it is impractical to develop written lockout/tagout procedures for all points on the premises wiring system. Instead, a generic procedure should be written requiring the lockout of all electrical sources feeding the equipment to be worked on. Such a generic procedure would be useless if not supplemented with resources that make it relatively simple to identify the sources of electricity to any point of the premises wiring system. Therefore, a one-line diagram of the premises wiring system and proper labeling of all electrical panels are essential to efficient implementation of the generic lockout/tagout procedure (see Figure 4.6).

Labels should list the electrical sources that supply a cabinet and the loads that the cabinet supplies. Where there are multiple sources of electricity that supply a cabinet, the location and voltage should be clearly indicated as well as a warning that multiple sources of electricity are present. When maintenance is scheduled, the diagrams, labels, and electrical schematics must be consulted to determine exactly what must be locked out and which loads the lockout will affect. The generic procedure must require that these resources be consulted to identify all energy sources for the cabinet. The generic procedure should also require that the maintenance supervisor review the specific lockout

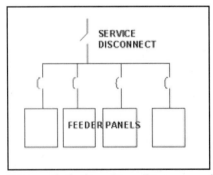

Figure 4.6: One-Line Diagram of Electrical Premises Wiring

steps on a case-by-case basis to ensure adequate protection. A second check of procedural steps has often proven to be absolutely essential to safeguard employees.

Policies versus Practicalities

As stated previously, a safety policy must be well considered, communicated, and enforced. An example of an electrical safety policy that is well-intentioned but may be impractical is to forbid work on live parts. In many cases, numerically controlled (NC) or computer numerically controlled (CNC) machines are designed without adequate consideration of lockout requirements. Many machines have a single electrical disconnect for drive power and control power. This is a good feature when loss of control power does not cause difficulties in restarting the machine. However, many machines lose their programming and knowledge of the tool's location when control power is interrupted. Such a machine is frequently worked on while energized in order to minimize down time and impact on production. It is essential to identify the presence of such machines in the facility before making policies that either cannot or will not be implemented. Once such machines are identified, a means can sometimes be found to lockout the drive power while maintaining control power. Modification to the machine's electrical circuitry is often required. Limitations on what tasks may be performed using an alternate lockout procedure must be determined and stated in the procedure's purpose.

Another situation that frequently is encountered is to feed multiple, unrelated equipment from a single energy isolation point. This type of installation is a violation of the NEC (i.e., Article 430-102) and creates situations where maintenance personnel could be pressured to work on live parts for the sake of production. Each piece of equipment that can conceivably be worked on independently of other equipment must have its own independent energy isolation source. Require that the production engineers, production managers, and maintenance manager concur in deciding what equipment can be safely worked on independent of other equipment.

Production engineers frequently design integrated machine systems for a particular process. The production engineer's perception frequently is that each piece of equipment in the system is essential and so cannot be worked on independent of other equipment in the system. The equipment is therefore not required to have independent energy isolation points. However, the production manager's perception is that, through ingenuity and manpower, a workable solution can be found such that a given piece of equipment could be placed out of service without placing the entire system out of service. Such decisions, if not designed into the system, create safety issues for maintenance personnel and for the operators implementing the workable solution. The maintenance manager should be able to provide insight into the difficulties of performing anticipated tasks, such as the space needed for dismantling equipment and access to the machine by fork trucks or cranes. Access and space requirements could make working on part of an operational system highly hazardous or even impossible. Finally, the safety professional must consider whether maintenance personnel can be safeguarded while performing work on part of an operational system, and whether operators can be safeguarded while working on out-of-service equipment. All these considerations affect the decision of how to safely provide energy to machines.

Some Lockout/Tagout Pitfalls

We have already discussed the need for lockout/tagout procedures and their requirements. Some common mistakes made when developing lockout/tagout procedures are:

Figure 4.7: Old-Style Electrical Disconnect

- Not explicitly stating the purpose of the procedure

- Not including all energy sources

- Rotating parts that may continue to rotate long after the machine is turned off

- Parts of the machine that are raised above their lowest point of travel

- Trapped hydraulic pressure between directional control valves and hydraulic cylinders

- Not providing adequate steps to release trapped pressure

- Not providing adequate steps to verify that the sources of energy have been isolated and released

Another common mistake is using a lockout/tagout procedure that was intended for maintenance of a machine to perform electrical maintenance inside the control cabinet. Depending on the circuitry and the type of electrical disconnect on the cabinet, the lockout/tagout procedure for the machine may not adequately protect employees working inside the control cabinet.

Additionally, some electrical disconnects that are capable of being locked do not prevent opening the enclosure door and thus bypassing the lockout. The type of electrical disconnect pictured in Figure 4.7 has a lockable handle mounted on the enclosure door. This old-style disconnect often does not prevent opening the door when the handle is locked. When the door is opened, the handle disengages from the energy isolation device. The disconnect switch can be placed in the ON position by using a pair of pliers. If this is the case, an additional lock on the enclosure door to keep it shut may be necessary.

Figure 4.8: New-Style Switch on Door

Another kind of lockable switch that may or may not be adequate is pictured in Figure 4.8. This type of switch can be an electrical disconnect where power to the cabinet passes through the switch. If the switch is designed as a disconnect with positive opening contacts, this is an adequate energy isolation device. If this switch controls power to a relay that opens and closes the prime mover power disconnect, then this switch is not adequate as an energy isolation device. When the switch is mounted on the door as in the picture below, its adequacy is questionable. The circuit should be reviewed to determine the switch's function and adequacy as an energy isolation device. The manufacturer's identification or labeling of the switch as an electrical lockout point does not necessarily mean it meets OSHA's standards or provides adequate protection for employees.

Working on or near Exposed Live Parts

Working in the vicinity of exposed live parts should only be allowed if absolutely necessary. This section discusses the requirements for premises wiring and does not apply to power generation, transmission, and distribution equipment. Similar considerations are required for such equipment, but the specific requirements differ significantly. For example, two qualified employees are required to work together for certain types of work on live lines. The approach boundaries are defined and presented in a different way. The approach boundaries for live lines of power generation, transmission, and distribution equipment are given separately for phase to ground and phase to phase exposures and include overvoltage considerations and altitude corrections. See 29 CFR 1910.269 and the NESC for specific guidance on working near and on exposed live parts of power generation, transmission, and distribution equipment.

OSHA in its standard, Selection and Use of Safe Work Practices (29 CFR 1910.333), permits working in the vicinity of exposed live parts under one of three conditions:

1. To de-energize the circuit would introduce additional or increased hazards.

2. To de-energize the circuit would be infeasible due to equipment design or operational limitations.

3. The circuit is less than 50 V and there will be no increased exposure to electrical burns or to explosion due to electric arcs.

The first condition is intended to include interruption of life support equipment, deactivation of emergency alarm systems, shutdown of hazardous location ventilation equipment, or removal of illumination for an area.

The second condition is intended to include voltage and current testing of electric circuits for troubleshooting and verification tests. Another example of this condition would be work on circuits that form an integral part of a continuous chemical process that would otherwise require the entire process to be shut down. There is sometimes difficulty in determining whether interruption to a specific process would be inconvenient or infeasible. The determining factor is usually the frequency of shutdown or interruption of the process. Continuous processes in chemical plants, such as the example given in the OSHA standard, typically run uninterrupted for long periods of time, even years. If a continuous process is typically shut down daily or weekly, it should be shut down rather than require work on exposed live parts. The design of processes to allow lockout of individual pieces of equipment is essential to minimizing the need to work on exposed live parts.

If one must work near exposed live parts, it is essential that precautions be taken to protect the employee from electrical hazards. Certain routine and periodic tasks should be described through a safe work procedure. Such procedures might include changing fluorescent light ballasts while the circuits are live, taking voltage measurements while troubleshooting in programmable logic control cabinets, operating high voltage electrical switches, or working above crane bridge rails. Such procedures are discussed in NFPA 70E, Part II and the NESC. The rest of this section discusses some elements of those procedures that must be considered. In developing safe work procedures, remember that they must be consistent and compliant with the electrical safety policies and the electrical safety management strategy already in place.

Work Plans

A work plan is one method that can be used to administratively ensure that the hazards of a task are thoroughly considered before work begins. A work plan is recommended for any work inside the restricted approach boundary. Such a plan might be a standard operating procedure for routine tasks, or the plan may have to be developed on a case-by-case basis. Where the work entails crossing the prohibited approach boundary, the work plan should include the reason such work is required and cannot be performed with the parts de-energized. Additionally, a hazard/risk evaluation should be performed when the work entails crossing the prohibited approach boundary (see Chapter 3). The work plan should be documented so that there is no chance for miscommunication and so that the plan can be evaluated later for improvement when the task must be performed again. It is not necessary that the plan be formal or of any specific format; however, it should be thorough and contain the following elements:

- Purpose of task
- Description of work
- Limits of approach for the voltages involved
- Safeguards to be used
- Qualification and number of employees needed

Limits of Approach to Exposed Live Parts

The risk to which an employee is exposed is dependent on his/her proximity to the electrical hazards and on the voltage of the exposed live parts. A series of concentric boundaries are defined around exposed live parts to categorize the precautions necessary to protect employees crossing those boundaries.

The Flash Protection Boundary is the boundary within which there is a danger from arc flash. Arc flash protection is required when inside this boundary.

The Limited Approach Boundary is the boundary within which unqualified employees are not permitted without qualified supervision. Qualified supervision is supervision qualified to work near exposed live parts. Additionally, the unqualified employee must be competent to perform the assigned task. Working within this boundary is considered working near exposed live parts.

The Restricted Approach Boundary is the boundary within which unqualified personnel are not permitted under any circumstances.

The Prohibited Approach Boundary is the boundary within which qualified personnel are considered working on exposed live parts. Qualified employees must be specifically trained and competent to work on exposed live parts.

According to NFPA 70E – 2000, Part II, the boundaries for exposed live parts are as follows:

Table 4.2: NFPA 70E, Part II, Table 2-1.3.4 Approach Boundaries to Live Parts for Shock Protection (All dimensions are distance from live part to employee.)

Nominal System Voltage Range, Phase to Phase	Limited Approach Boundary		Restricted Approach Boundary, Includes Inadvertent Movement Adder	Prohibited Approach Boundary
	Exposed Movable Conductor	Exposed Fixed Circuit Part		
0 to 50	Not specified	Not specified	Not specified	Not specified
51 to 300	10 ft 0 in.	3 ft 6 in.	Avoid contact	Avoid contact
301 to 750	10 ft 0 in.	3 ft 6 in.	1 ft 0 in.	0 ft 1 in.
751 to 15 kV	10 ft 0 in.	5 ft 0 in.	2 ft 2 in.	0 ft 7 in.
15.1 kV to 36 kV	10 ft 0 in.	6 ft 0 in.	2 ft 7 in.	0 ft 10 in.
36.1 kV to 46 kV	10 ft 0 in.	8 ft 0 in.	2 ft 9 in.	1 ft 5 in.
46.1 kV to 72.5 kV	10 ft 0 in.	8 ft 0 in.	3 ft 3 in.	2 ft 1 in.
72.6 kV to 121 kV	10 ft 8 in.	8 ft 0 in.	3 ft 5 in.	2 ft 8 in.
138 kV to 145 kV	11 ft 0 in.	10 ft 0 in.	3 ft 7 in.	3 ft 1 in.
161 kV to 169 kV	11 ft 8 in.	11 ft 8 in.	4 ft 0 in.	3 ft 6 in.
230 kV to 242 kV	13 ft 0 in.	13 ft 0 in.	5 ft 3 in.	4 ft 9 in.
345 kV to 362 kV	15 ft 4 in.	15 ft 4 in.	8 ft 6 in.	8 ft 0 in.
500 kV to 550 kV	19 ft 0 in.	19 ft 0 in.	11 ft 3 in.	10 ft 9 in.
765 kV to 800 kV	23 ft 9 in.	23 ft 9 in.	14 ft 11 in.	14 ft 5 in.

For single phase circuits, multiply the single phase voltage by the square root of 3 (i.e., 1.732) and enter the table above. Treat DC circuits as single phase circuits for the purposes of entering this table.

Working near Overhead Lines

Work near energized high voltage overhead lines is dangerous and has been the cause of many serious accidents. Unqualified employees should never approach overhead lines closer than 10 feet plus an additional 4 inches for every 10 kV over 50 kV. This approach boundary applies to all equipment and tools not specifically rated for the voltage involved.

One example of inadvertent encroachment of this approach boundary is driving fork trucks under overhead lines and stacking containers near the overhead lines. If this hazard is not identified and adequately controlled, it could lead to a fatality, serious accident, or fire.

Using long poles in the vicinity of overhead lines is another common way that personnel inadvertently encroach within the limited approach boundary. For example, an employee was standing on top of a vertical cylindrical tank near the property line. The tank was at least 20 feet tall. He was using a very long pole with a brush attached to the end to clean the inside of the vertical cylindrical tank. The employee was electrocuted when the long pole inadvertently came within the prohibited approach boundary of overhead lines at the edge of the property.

The potential for a similar accident exists where overhead lines provide service to the facility transformer and are located close to the building. Employees working on the roof may inadvertently encroach on the limited approach boundary, especially if using long poles to rake gravel, remove snow, or lay tar. If employees or contractors were to get too close to overhead lines, they could be seriously injured or killed.

If work is to be performed near overhead lines, the lines must be de-energized and grounded, or otherwise guarded by the controlling organization (e.g., electric company). Protective measures include guarding or insulating. If such measures are used, they must prevent employees from contacting overhead lines directly or indirectly through conductive materials, tools, or equipment. (See 29 CFR 1910.333(c)(3) "Overhead lines.")

The work practices used by qualified persons installing insulating devices on overhead power transmission or distribution lines are covered by 29 CFR 1910.269, not by 29 CFR 1910, Subpart S—Electrical. Qualified employees are required to stay outside the restricted approach boundary of energized overhead lines under Subpart S.

Flash Hazard Analysis

The flash protection boundary is given for various three-phase voltages in NFPA 70E, Part II, Table 2-1.3.4 (see Table 4.2). The flash protection boundary can be greater than the limited approach boundary for low voltages. The distances depend on voltage and, for voltages less than 600V, are based on typical values of 50kV of available fault current and a time to clear a fault of 0.1 seconds. The table entries are sufficient for most industrial plant applications. For unusual circumstances or voltages not included in the table, a calculation should be made under competent engineering supervision to determine the flash protection boundary. See NFPA 70E, 2-1.3.3.2 for a discussion of this equation and its limitations.

Insulated Tools

Insulated tools and handling equipment are required if the tools or equipment might make contact with exposed live parts. Protective coverings on the insulated tools are required if the insulating capability of the tools can be damaged.

Fuse handling equipment, insulated for the circuit voltage, is required to remove or install fuses when the fuse terminals are energized.

Ropes and handlines used near exposed live parts must be nonconductive.

Protective shields, barriers, or insulating materials are required to protect employees from electrical hazards whenever there is a chance of arcing, dangerous electrical heating, or accidental contact with exposed live parts. Such materials can be used as an alternative to electrical PPE where the task to be performed makes wearing PPE impractical. Ensure employees are completely protected from electrical hazards while installing protective insulating materials.

Alerting Techniques

Unqualified people must be protected from inadvertent contact of exposed live parts during servicing or maintenance. The use of signs and barricades is required whenever access to exposed live parts by unqualified people is afforded because of servicing or maintenance. Barricades may be barriers, caution tape, or cones. If access cannot be adequately guarded, an attendant is required to prevent unqualified persons from gaining access to exposed live parts.

Chapter Five

Electrical Safety Training

Overview

Safety policies, procedures, and work practices are not sufficient safeguards in themselves. Once safe work procedures and safety rules are developed, it is essential that all personnel know the rules and procedures that directly affect them and have a basic awareness of electrical safety. Training is a critical part of managing electrical safety. Without effective training, there is no communication of the policies, procedures, and management expectations of safe work practices.

Training is a vital part of any safety program and consists of two parts: providing information to the trainees, and receiving confirmation from the trainees that they understand the information. Training can be classroom, on-the-job, or self-paced (e.g., book, computer, or online). Confirmation of understanding can be by oral or written test, interview, or observation of performance.

Identifying Training Needs

The JHA is a good source of training material. It is also a good source for identifying needed training content. Training must be part of all hazard controls that are administrative in nature. (Administrative controls are controls other than engineering controls, and include safe work practices or procedures, job rotation, work/rest cycles, and PPE.) Additional documents that point to needed training content include accident/incident investigation reports, safety inspection reports, employee suggestions, and safety committee recommendations. Do not forget that regulations and standards indicate much that must be included in training.

Once the needed training content is identified, the question becomes who must be trained. While not all employees need to receive the same level of training, all employees, whether in a production setting or an office, can benefit from electrical safety awareness training. Employees working around electrical equipment in an industrial setting that are not permitted to expose themselves to electrical hazards without direct supervision are unqualified employees. These employees need additional information about the hazards in their workplace and the controls on those hazards—what the employee can do and what the employee is prohibited from doing. And of course employees that are qualified to work on or near exposed live parts must be trained how to perform those duties safely.

This training, known as refresher training, can be scheduled at some point during the year. But new hires, temporary employees, and even contractors on the premises require training too. These trainees come into the facility throughout the year and may not be able to wait till the next scheduled training. It is essential that all employees, whether permanent or temporary, have the training they need to perform their tasks safely before they are exposed to the hazards of the job. Therefore, one-on-one training sessions may be needed to give new hires and temporary employees the information they need. Ensure that the hiring/orientation process includes verification that all required training pertinent to the employee's job is accomplished before the employee is assigned duties in the workplace. Where a temporary agency or contractor is used to supply personnel for a project or on a semi-permanent basis, the responsibility to conduct training can be delegated to the agency/contractor. They should conduct training prior to sending personnel to the facility for work. However, the facility should ensure that any temporary employee or contractor has the required training to work safely in the facility. The extent to which the facility should verify proper training depends on the length of the employee's employment at the facility, the scope of the work, the direction/supervision provided at the facility, and the risk of the task performed. Even if the contractor will perform a specific project without direct supervision or direction from the facility, the facility management should make sure the contractor will not take unnecessary risks. The contractor selection process should ensure that a contractor has the proper certifications, education, licenses, reputation, and safety record. After a contractor is hired, the facility supervisor/manager responsible for the work being done should observe contractor activities and ensure compliance with safe work practices and policies.

Conducting Training

Training can be conducted in a variety of formats depending on the content and the time it takes to convey the information.

- Reading pamphlets or flyers
- Listening directly to a teacher, as in a classroom
- Watching a video or other electronic media
- Practicing tasks and on-the-job training

Regardless of the format of training, the employee must always be able to ask questions and receive timely responses. A video or self-paced training does not afford opportunities for questions and discussion without being supplemented by other communication means. Without the ability to ask questions and get timely responses, the training is not considered adequate by OSHA.

Of course, training and responses to questions must be in a language that employees understand. And the effectiveness of training is greatly increased when it is conducted in the employee's and instructor's primary language. When conducted in a secondary language, it is possible for the information to be misunderstood or communicated incorrectly, without it being at first apparent. Whenever possible, training materials should be translated into the primary language of the employee. Do not overlook photographs, graphics, videos, or hands-on examples. Even if the trainees understand English, it is a good idea to have an interpreter available. This should be a bilingual employee fluent in both English and the second language. If a bilingual employee is not available, use the employee most fluent in both languages as an interpreter if needed to ensure understanding. If understanding is still not possible, a professional interpreter will probably be necessary.

When conducting the training, the trainer should provide an overview of the material, follow with details, and conclude with a summary. The trainer should always try to relate the information to the trainees' personal and professional goals, self-interests, and experiences. Pertinent incidents, accidents, and issues in the employees' workplace can be used as examples. The more they are familiar with the events, the better the example.

Verification of Understanding

After conducting training, the employer needs to determine if the training was effective. The best way to accomplish this is to establish measurable goals and objectives specific to the training content, then determine if these objectives have been met. The objectives should identify the following:

- What precisely has to be demonstrated
- How will the objective be demonstrated
- What constitutes acceptable demonstration

The goal of training is to provide the knowledge and/or skills that are critical to employees performing their duties safely. The goal of verification is to ensure employees have acquired the necessary knowledge and skills. It is necessary for the safety program that the knowledge and skills are retained.

Verification can be accomplished through written test, oral questions, or direct observation by supervisors, peers, or lead employees. The means of verification is dependent on the knowledge or skill being verified. Basic knowledge can be verified by asking simple true/false questions or fill-in-the-blank questions on a written test. A written test can be on paper or on a computer network. More complicated knowledge is better verified through oral discussion or direct observation. Skills are best verified by observing the employee using them properly. The proper donning (putting on), doffing (taking off), and wearing of PPE is a required demonstration for all types of PPE.

Program Improvement

Based on the verification results and on observed practices, incidents, and accidents that follow, the training may need to be improved. The format of training may not have been the ideal way to convey the critical information. Language barriers may persist. Employees may not have understood the material, participated in voicing questions, or discussed confusing information. If verification results show the training was successful, but subsequent events show it was not, the verification process should be improved. Even if the training is a success, there is always room for improvement. The trainees should have the opportunity to give suggestions and feedback about the training so that it can be improved upon. Remember that training is a part of the dynamic program model. The process is a repeating cycle with continuous improvement.

Required Training Content

The content of electrical safety training and other regulatory requirements can be found in the following OSHA standards:

- For premises wiring – 29 CFR 1910.332

- For electrical power generation, transmission, and distribution – 29 CFR 1910.269

Employees fall into two training categories – those employees qualified to work on or near of exposed live parts (i.e., qualified employees) and those employees that are not so qualified (i.e., unqualified employees).

All Employees

Everyone should have a general awareness of the hazards of electricity and the limitations to their responsibilities. Office employees often work with cord-and-plug equipment or appliances. Some office employees might attempt to reset circuit breakers and replace light bulbs, depending on the availability of maintenance personnel and the frequency of the condition. Operators on the production floor frequently work in the vicinity of open electrical panels. They quickly learn how to access critical components that require frequent adjustment or resetting. It is essential that all employees know what they are allowed to do and what is not permitted.

According to OSHA (29 CFR 1910.334), everyone should know the safety rules provided on the following checklist.

General Electrical Safety Rules

❏ Do not use flexible cords for raising or lowering the equipment.

❏ Do not fasten flexible cords with staples or otherwise hang them in a way that could damage the outer jacket or insulation.

❏ Visually inspect portable cord and plug equipment and extension cords before plugging them in. Look for external defects (such as loose parts, deformed and missing pins, damage to outer jacket, pinched insulation, or crushed outer jacket).

❏ Remove defective or damaged portable equipment and extension cords from service until repaired and tested to be safe.

❏ Ensure that plugs and receptacles are of proper mating configuration before attempting to plug in equipment.

❑ Only use three-wire extension cords (three-prong plug) with equipment that has a three-prong plug.

❑ Do not connect or alter attachment plugs and receptacles in a way that would prevent proper continuity of the equipment grounding conductor.

❑ Do not alter plugs to allow the grounding pole to be inserted into slots intended for connection to current-carrying conductors.

❑ Do not use adapters that interrupt the continuity of the equipment grounding connection.

❑ Only use equipment approved for wet locations in highly conductive work locations (such a those inundated with water or other conductive liquids) or in job locations where employees are likely to contact water or conductive liquids.

❑ Ensure hands are dry when plugging and unplugging flexible cords and cord and plug connected equipment, if energized equipment is involved.

❑ Use insulated gloves to handle wet plugs and receptacles that are energized.

❑ Ensure locking type connectors are properly secured after connection.

❑ Do not use non-load-break-type cable connectors, fuses, terminal lugs, or cable splice connections as routine disconnecting means.

❑ Do not re-energize a circuit after a circuit protective device has opened until it has been determined that the equipment and circuit can be safely energized. When it can be determined from the design of the circuit and the involved overcurrent devices that the automatic operation of a device was caused by an overload rather than a fault condition, no examination of the circuit or connected equipment is needed before the circuit is re-energized.

❑ Do not modify overcurrent protection of circuits and conductors.

❑ Only qualified persons may perform testing work on electric circuits or equipment.

❑ Do not use electrical equipment capable of igniting flammable liquids, vapors, or gases, combustible dusts, or ignitable fibers or flyings when introducing these materials into the area.

Employees Not Qualified to Work near Exposed Live Parts

The federal OSHA standard for electrical safety training for premises wiring is 29 CFR 1910.332, Training. The standard requires that personnel with a significant risk of exposure to electrical hazards receive training on how to avoid or protect themselves from those hazards. Such personnel include:

- Blue collar supervisors
- Electrical and electronic equipment assemblers
- Electricians
- Material handling equipment operators
- Painters
- Stationary engineers
- Electrical and electronic engineers
- Electrical and electronic technicians
- Industrial machine operators
- Mechanics and repairers
- Riggers and roustabouts
- Welders

According to OSHA (29 CFR 1910.333), employees who are not qualified to work near or on exposed live parts but who may be required to enter the limited approach boundary with qualified supervision must know proper safety rules and be familiar with alerting techniques that qualified employees might use. Alerting techniques are used to warn and protect employees from hazards that could cause injury due to electric shock, burns, or failure of electric equipment parts. The OSHA requirements come from 29 CFR 1910.333 and 335. The checklist provided in Figure 5.1 provides guidelines for employees not qualified to work near exposed live parts.

Figure 5.1: Guidelines for Employees Not Qualified to Work near Exposed Live Parts

❑   Do not enter spaces containing exposed energized parts, without adequate illumination.

❑   Do not perform tasks near exposed live parts where you cannot see what you are doing.

❑   Use protective shields, barriers, or insulating materials as necessary to avoid inadvertent contact with exposed live parts in a confined or enclosed space.

❑   Secure doors and hinged panels to prevent their swinging into an employee and causing the employee to contact exposed energized parts.

❑   Ensure conductive materials and equipment do not contact exposed energized conductors or circuit parts. Follow safe work practices and procedures for using long conductive objects in proximity to exposed live parts.

❑   Use only nonconductive portable ladders where there is a potential to contact exposed live parts.

❑   Do not wear conductive articles of jewelry and clothing if they might contact exposed energized parts.

❑   Do not perform housekeeping duties near exposed live parts without precautions to prevent electrical contact.

❑   Do not use conductive cleaning materials (steel wool, metalized cloth, silicon carbide, or conductive liquid solutions) close to energized parts without precautions to prevent electrical contact.

❑   Do not defeat electrical interlocks. Only a qualified employees may defeat an electrical safety interlock temporarily while he or she is working on the equipment.

❑   When working near exposed live parts, use insulated tools or handling equipment. Protect the insulating material of the insulated tools or handling equipment if it is subject to damage.

❑   Use fuse-handling equipment, insulated for the circuit voltage, to remove or install fuses when the fuse terminals are energized.

❑   Use nonconductive ropes and handlines near exposed energized parts.

❑   Guard exposed live parts to protect unqualified persons from contact with the live parts during maintenance.

❑   Know what safety signs, safety symbols, or accident prevention tags look like and mean.

❑   Know that nonconductive barricades are there to protect against electrical shock and arc flash and should never be circumvented.

❑   Know that protective shields, protective barriers, and insulating materials protect against shock and burns and should not be circumvented or moved by unqualified employees.

Employees Qualified to Work on or near Exposed Live Parts

Employees qualified to work on or near exposed live parts must know how to perform their duties safely. The following topics must be covered and depend greatly on the specific electrical hazards present in the workplace and to a lesser extent which standards and regulations apply (i.e., premises wiring or power generation, transmission, and distribution equipment):

- How to distinguish exposed live parts from other parts of electric equipment
- Approach distances specified in NFPA 70E, Part II, Table 2-1.3.4 (see Table 4.2) and the corresponding voltages to which the qualified person will be exposed
- How to determine the nominal voltage of exposed live parts
- How to determine the degree and extent of the hazard and the personal protective equipment and job planning necessary to perform the task safely
- How to visually inspect test instruments, equipment, and all associated test leads, cables, power cords, probes, and connectors before use
- What to do if a defect or evidence of damage of test instruments and equipment are discovered
- How to ensure test instruments, equipment, and their accessories are properly rated for the intended voltage and for the intended environment

Working on exposed live parts (i.e., inside the prohibited approach boundary) is much more dangerous than working near exposed live parts. If qualified employees will work on exposed live parts, they must be trained on the following:

- Special precautionary techniques
- Personal protective equipment
- Insulating and shielding materials
- Insulated tools

If qualified employees work on exposed live parts and it is reasonable to require them to provide emergency response to electrical accidents where they are working, they must also be trained on the following:

- Methods of release of victims from contact with exposed energized parts
- First aid and emergency procedures
- Methods of resuscitation (i.e., CPR)

Supervisors of Qualified Employees

Supervisors of qualified employees should have the same training on safe work practices as the qualified employees. The supervisors will review safe work plans, lockout procedures, PPE requirements, and risk analyses. They need to know the safe electrical practices, standards, and regulations in order to ensure adequate steps are taken to protect employees and minimize the company's risk and liability. When developing a list of attendees for training, include the supervisors, their backups, and engineering support personnel who may find it necessary to approach exposed live parts within the restricted approach boundary.

Chapter Six

Safety Management Review of Electrical Systems

Overview

The potential hazards of electrical systems are great. Improperly designed electrical systems can lead to overheated components or conductors and subsequent fire. Improperly installed electrical systems can lead to electrical shock of operators or arcing resulting in fire. It is extremely important that the design of electrical systems be critically reviewed for safety and compliance with codes. Once the electrical safety program is in place, a management system must be created to ensure that all new electrical installations or modifications do not impose additional risk, require new or modified safe work procedures, or require additional electrical PPE. Appropriate personnel should review plans for major modifications and new installations to determine if the proposed equipment change:

- Is compliant with electrical codes and standards, as well as other safety standards
- Is in agreement with the electrical safety strategy, policies, and existing safe work procedures
- Has complete and efficient means of lockout and any required machine-specific lockout procedures
- Will require additional electrical safe work procedures
- Will require additional insulated tools, measurement equipment, or PPE

Management of Change

Management of change is a management practice that captures proposed changes—significant modifications or new equipment—so that their safety can be assured prior to expending any funds toward making the change. There is nothing more frustrating to both safety professionals and product engineers than to install new equipment only to have someone say it cannot or should not be used because it is unsafe.

After a project is envisioned, developed, planned, specified, and constructed, it is too late to effectively ensure the safety of its operation. The safety of industrial equipment and its electrical installation must be designed into the equipment from the beginning. Therefore, it is necessary that the design undergo periodic safety review throughout the design process. A formal design process that incorporates safety will not happen without prior planning, the insistence of the safety professional, and the staunch support of upper management.

If there is no formal design process that incorporates safety review at your facility, upper management must be convinced of the need for such a management system. The following train of thought is provided to assist the safety professional in convincing upper management of the need for safety review during project planning.

When a requirement is omitted from the design of a system, the consequences can be failure of the system to meet expectations. Every system has inherent risk from hazards that are an integral part of the system. If the design does not comply with safety requirements or has unacceptable inherent risk, the system may cause unexpected or unanticipated injury or damage. If any of these consequences are realized, the cost to the company could be catastrophic.

The cost to change the design after the system is constructed is much greater than the cost of incorporating the change into the planning phase. Therefore, it is most efficient to ensure compliance with safety requirements and an acceptable level of risk during the planning phase of the project.

A lot of effort is expended in making systems safe, developing procedures, and training employees. All this effort can be undone by making a significant modification to the equipment. Therefore, each proposed change must receive approval from a person in authority. This person is responsible for ensuring the proposed change will be safe and that all other safety program elements are updated accordingly before the change is implemented. He/she must be held accountable for accomplishing this vital task. The safety professional's job is to assist the responsible person in authority to ensure safety. The safety professional's expertise and familiarity with regulations and standards is essential to the success of management of change. The change should be reviewed for safety periodically throughout the project—during the conceptual phase, design phase, manufacturing/installation phase, and before the equipment is made operational.

Incorporating Safety Review into Project Planning

Often the design of an electrical installation changes during the course of construction. Changes in equipment design, layout, or other design considerations force changes to the design of the electrical installation. Often these changes occur after the specification of the electrical system has been finalized. It is essential that these changes be reviewed for their impact on safety. Therefore, a management system requiring safety review of design changes should be developed and implemented throughout the project.

Safety review can begin in the early stages of the project development (i.e., the conceptual phase). This preliminary hazard analysis should identify inherent hazards in the proposed change and determine the types of controls to be used for each hazard. These controls can be incorporated into the design during the next phase.

In the design phase, the concept may undergo many iterations of change. Each change should be reviewed to identify any new safety issues that may have been created by those changes. When the design is considered at least 70% complete and firm, a critical design review should be conducted. This review should be a detailed look at the drawings, plans, and specifications of the design to identify any safety issues and get them resolved while the design is on paper only.

In the next phase, the equipment is manufactured or purchased and assembled/installed. There are several opportunities for safety review during this phase. When equipment is received it should be inspected to ensure it meets plans, drawings, and specifications. This is an essential task to the overall safety of the project and should be conducted by the project manager. The adequacy of work plans and the enforcement of safe work practices during this phase will be an ongoing effort that should involve all levels of management, the employees performing the work, the safety committee, and the safety professional. Near the end of this phase, there should be one final and formal safety review prior to energizing the equipment (even for the first time during assembly).

Safety Review Prior to Power-On

Prior to powering up the circuit for the first time, the electrical installation should be carefully tested and inspected. A qualified employee should test the electrical continuity of each phase, neutral, and ground path of the electrical installation. This test will find open circuits along the installation. Where conductors are subject to movement or vibration, the conductors should be moved by hand while the continuity is tested to ensure loose connections are identified. Next, the qualified employee should test the resistance between each phase, neutral, and ground of the electrical installation. This test will identify any short circuits in the installation. Finally, the qualified employee should check that multiphase equipment has been connected for proper phase rotation. Perform all of these tests prior to powering up any part of the electrical installation.

Next the installation should be inspected for compliance with the NEC and the design specifications and plans. A competent supervisor should perform this inspection with the safety professional. The safety professional should ensure that adequate attention was given by the inspector. Use a checklist to document the various installation characteristics that have been specified for the installation and required by the NEC for the method of electrical installation used. The safety professional and competent supervisor should ensure that the tests discussed above have been performed successfully.

If the electrical installation passes these tests and inspection, the electrical installation may be energized. This safety review prior to power on should be documented and include the following points:

- Electrical installation meets the NEC.
- Machine wiring meets NFPA 79.
- Machine point of operation, power transmission apparatus, and other moving equipment are adequately safeguarded.

Final Safety Review

After the acceptance tests have been performed satisfactorily, but before the equipment is placed into operation, the equipment should receive one last safety review and formal sign-off. Placed into operation means operating the equipment in a production mode. Making trial parts for quality certification is production-mode operation. This safety sign-off should include all aspects of safety and health, including:

- Adequate access to the workspace and emergency equipment required to safely operate the machinery is provided (at least 18 inches wide access to workstation).

- There is adequate illumination and ventilation.

- There are adequate lockout/tagout procedures.

- A Job Hazard Analysis for any new task has been completed.

- A confined space entry procedure has been developed and any confined spaces have been added to the list of confined spaces, if needed.

- Operating and maintenance manuals are available.

- The prevented maintenance system has been updated to include the equipment.

- Operators have been trained how to safely operate the equipment, about the hazards and safety features of the equipment, and about any PPE required to operate the equipment.

The safety review prior to power on and the final safety sign-off can be documented on forms similar to the ones shown in Figures 6.1 and 6.2.

The installation method is in accordance with all applicable provisions of the National Electrical Code	
There is required clearance in front of all electrical panels that might have to be opened while energized.	
The electrical disconnect is readily identifiable or else labeled as to its function and is lockable.	
All sources of electrical power to switchboards or panelboards are identified and labeled on the equipment.	
The nominal voltage of all panels is labeled.	
Controls are clearly labeled as to their function.	
Start/cycle/on controls are green and protected from inadvertent activation.	
Stop controls are red and are not protected from inadvertent activation.	
Emergency stop controls are red, mushroom shaped, accessible from each operator workstation, and not protected from inadvertent activation.	
Point of operation, power transmission apparatus, and other moving parts of equipment are adequately safeguarded.	
Interlocks and safeguarding devices do not rely on programmable logic to accomplish the safety function but are hardwired into the circuit.	
There is adequate illumination, temporary or permanent.	
There is adequate ventilation, temporary or permanent.	
There is adequate access to workstations and equipment (at least 18 inches wide and 6 feet 6 inches high).	

Appropriate signature block

Figure 6.1: Safety Review Prior to Power On

There is an adequate lockout/tagout procedure.	
A Job Hazard Analysis and PPE assessment has been completed for any new equipment.	
Any confined spaces have been identified, added to the facility list, labeled, communicated, and adequate entry procedures developed.	
Operating and maintenance manuals are available.	
There are adequate operating procedures available.	
The preventive maintenance system has been updated to include any equipment changes.	
Operators have been adequately trained how to safely operate the equipment, about the hazards and safety features of the equipment, and about any needed PPE.	
The equipment acceptance tests are all satisfactory.	

Appropriate signature block

Figure 6.2: Final Safety Sign-Off

Safety and Maintainability in Design

Several characteristics of electrical installation design require early consideration in the design development process. These include adequate clearance around electrical installations, choosing an appropriate installation method, and maintainability. Maintainability is the characteristic of a design permitting efficient maintenance to be performed. If maintenance cannot be performed quickly, down time will be increased—reducing productivity. If maintenance cannot be performed at a safe location or while the system is in a safe state, unnecessary risk is incurred. These and other characteristics of the design must be reviewed from multiple perspectives: productivity, safety, maintainability, reliability, feasibility, and cost.

Planning for Adequate Clearance

One of the most common electrical compliance issues cited by OSHA is failure to provide required clearance around electrical components and enclosures. OSHA and the NEC require specific clearances around electrical components that are likely to require access while energized.

Hinged doors on electrical enclosures that are likely to be opened while energized must be able to be opened at least 90° (see Figure 6.3). The entire width of the enclosure must be unobstructed. For enclosures of parts not more than 600 V, there must be at least 30 inches of unobstructed width in front of the enclosure. (For enclosure of parts more than 600 V, there must be at least 36 inches of unobstructed width in front of the enclosure.) This width does not have to be centered on the enclosure opening as long as the doors can open 90°.

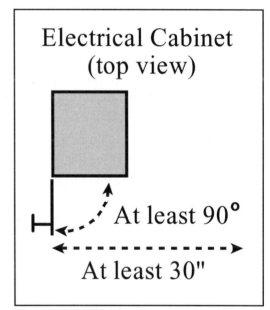

Figure 6.3: Width of Access

Access to the enclosure must be unobstructed from the floor or work platform to a height of at least 6 feet 6 inches above the floor or platform according to the latest edition of the NEC. OSHA only requires 6 feet 3 inches for enclosures of parts less than or equal to 600 V.

Access must be clear in front of the enclosure for a distance depending on the conductive quality of the material opposite the enclosure opening (see Figure 6.4). The required clearance is measured from the enclosure door.

An insulated object or material includes drywall, wood paneling, or plywood.

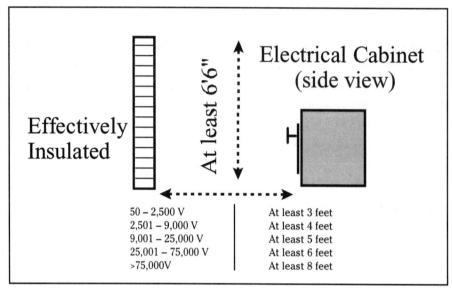

Figure 6.4: Clearance in Front of Electrical Enclosures Opening toward Insulated Parts or Materials

If the enclosure opening faces a grounded object or material, the required clearance in front of the enclosure is increased (see Figure 6.5). A grounded object or material includes concrete, masonry, corrugated metal, I-beams, or grounded electrical equipment.

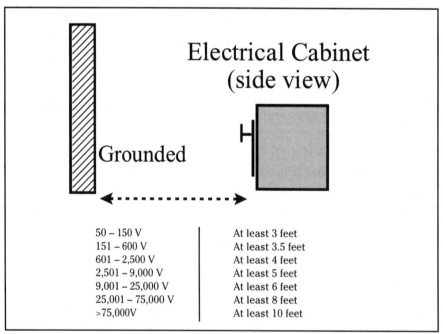

Figure 6.5: Clearance in Front of Electrical Enclosures Opening toward Grounded Parts

If the enclosure opening faces exposed live electrical parts, the required clearance in front of the enclosure is increased again (see Figure 6.6). Exposed live electrical parts include circuits inside electrical enclosures that open facing each other.

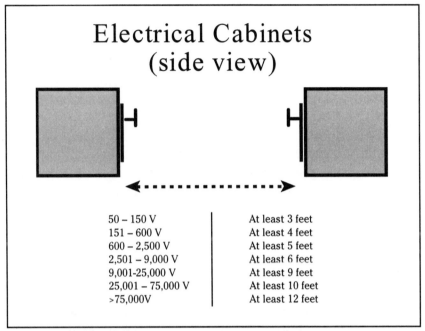

Figure 6.6: Clearance in Front of Electrical Enclosures Opening toward Exposed Live Parts

For enclosures containing exposed live parts not more than 600 V, the NEC allows the distance to be reduced to 42 inches in industrial establishments where (1) access to exposed live parts is only permitted to qualified employees and (2) there is a written policy that prohibits both cabinets from being opened at the same time while energized. The OSHA regulations do not permit this reduction in clearance. However, OSHA recognizes the NEC as an industry consensus standard describing good engineering practice. Where compliance with a more recent industry consensus standard provides an equivalent level of safety, OSHA considers violations of the federal standard to be de minimus—meaning no fine would be involved.

Choosing an Appropriate Installation Method

The method of installation is best chosen by production engineering or electrical maintenance personnel. The safety professional does not have to have an intimate knowledge of electrical installation methods or the NEC. However, the safety professional must ask whether the chosen method is appropriate given the intended circumstances. The NEC Chapter 3 provides detailed guidance and requirements for each installation method. Each electrical installation method is discussed in a separate article. Of critical interest to the safety professional are the paragraphs within each article on "Definition," "Uses Permitted," and "Uses Not Permitted." These paragraphs should be consulted to verify that the installation method chosen is appropriate for the intended circumstances.

Following the initial paragraphs in each article on electrical installation methods are the installation requirements including support, minimum bend radius, and other safety concerns. The safety professional may not be qualified to inspect the electrical installations to confirm compliance with the NEC, but he/she can require that the NEC be consulted, considered, and complied with by the personnel responsible for the electrical installation.

Fixed wiring methods are preferable to flexible cables unless the electrical conductors must be flexible. In general, rigid wiring methods are preferable to fixed wiring methods using flexible conduit or tubing. All of the following fixed installation methods are described in the NEC. The following is

> **The preferred method of installation of electrical conductors is the one that provides the most physical protection for the conductors and the best insulation of the conductors.**

a short description of methods to aid the safety professional in classifying the types of methods used in the facility.

Underground Installations (ARTICLE 300-5)

Underground installations include direct burial in earth such as underground feeder and branch-circuit cable, or cable run through buried raceways. Direct buried cable is specially listed cable with a covering that is flame retardant; moisture, fungus, and corrosion resistant; and suitable for direct burial in the earth.

Cable Trays (ARTICLE 318)

A cable tray is a unit or assembly forming a structural system used to securely fasten or support cables and raceways.

Cable trays can be ladder-type, ventilated trough, ventilated channel, solid bottom, or other similar support structure. Cables and single conductors within limits are laid on these cable trays. Cable trays are frequently encountered in industrial and commercial environments for power feeder and

> **A raceway is an enclosed channel of metal or nonmetallic material designed expressly for holding wires, cables, or busbars.**

branch circuits, and for data/instrument cable distribution. Cable tray systems cannot be used in hoistways or where subject to severe physical damage.

Open Wiring on Insulators (ARTICLE 320)

Open wiring on insulators is an exposed wiring method using cleats, knobs, tubes, and flexible tubing for the protection and support of single insulated conductors run in or on buildings and not concealed by the building structure. Open wiring on insulators is permitted for 600 volts, nominal, or less, and only for industrial or agricultural establishments, indoors or outdoors, in wet or dry locations, where subject to corrosive vapors, and for service connections.

Messenger Supported Wiring (ARTICLE 321)

Messenger supported wiring is an exposed wiring support system using a messenger wire to support insulated cables. This method can be used for service drops, feeders, and branch circuits. An example with which most people are familiar is the utility connection from an electric pole to the roof of a house or building.

Concealed Knob-and-Tube Wiring (ARTICLE 324)

Concealed knob-and-tube wiring is a wiring method using knobs, tubes, and flexible nonmetallic tubing for the protection and support of single insulated conductors. This wiring method allows the wiring to be concealed by the building structure. It is only permitted for extensions of existing installations unless special permission from the local building code official is obtained. It is not permitted to be used in hollow spaces of walls, ceilings, and attics where enveloped by loose, rolled, or foamed-in-place thermal insulating material.

Electrical Nonmetallic Tubing (ARTICLE 331)

Electrical nonmetallic tubing is a listed raceway of circular cross section made of pliable, nonmetallic, corrugated material. It is resistant to moisture and chemical atmospheres and is flame retardant. A pliable raceway can be bent by hand with reasonable force. It is used for many purposes including chemical environments and wet locations. It is typically not permitted in hazardous (classified) locations or theaters and similar locations. It cannot be used for:

- The support of fixtures and other equipment
- Where subject to ambient temperatures in excess of 122ºF unless listed otherwise
- For conductors whose insulation temperature limitations would exceed those for which the tubing is listed
- Buried underground
- For circuits over 600 V
- Where exposed to the direct rays of the sun unless identified as sunlight resistant.

Intermediate Metal Conduit (ARTICLE 345)

Intermediate metal conduit is a listed steel raceway of circular cross section, similar to rigid metallic conduit but of thinner-walled construction. It can be used in the same way as rigid metallic conduit with the exception that the thinner wall does not provide the same corrosion resistance as rigid metallic conduit.

Rigid Metal Conduit (ARTICLE 346)

Rigid metal conduit is a listed metal raceway of circular cross section with integral or associated couplings, approved for the installation of electrical conductors and used with listed fittings to provide electrical continuity. Basically it is like putting conductors in a rigid pipe. This wiring method can be used practically anywhere with special provisions for the hazards and environment.

Rigid Nonmetallic Conduit (ARTICLE 347)

Rigid non-metallic conduit is a listed non-metallic raceway of circular cross section that is resistant to moisture and chemical atmospheres. This type of conduit is made of PVC, fiberglass, or similar materials. The use of rigid nonmetallic conduit is permitted to be concealed in walls, floors, and ceilings, buried, or installed above ground. It is useful in corrosive, wet, or damp locations. It is not permitted in hazardous classified locations, where ambient temperatures are above 122ºF unless listed otherwise. Nor is it to be used for conductors whose insulation temperature limitations exceed the conduit's temperature limitations, nor in theaters and similar locations.

Electrical Metallic Tubing (ARTICLE 348)

Electrical metallic tubing is a listed raceway made of pliable metallic tubing of circular cross section. It can be used in exposed spaces or concealed locations, buried underground or in concrete, or in corrosive environments where corrosion protection is provided. It is typically not used in hazardous (classified) locations or buried in cinder concrete or cinder fill unless otherwise protected. It cannot be used where subject to severe physical damage, where protected from corrosion solely by enamel, or for the support of fixtures or other equipment.

Flexible Metallic Tubing (ARTICLE 349)

Flexible metallic tubing is a listed raceway tubing that is circular in cross section, very flexible, metallic, and liquidtight without a nonmetallic jacket. Flexible metallic tubing is used for branch circuits in dry locations, concealed or accessible, for circuits of 1000 volts maximum. A typical use for flexible metallic tubing is inside air plenums or ducts. It is limited to 6 feet in length. It is not typically used in hazardous (classified) locations. It cannot be used in hoistways, in storage battery rooms, underground, embedded in poured concrete or aggregate, where subject to physical damage, or in lengths over 6 feet.

Flexible Metal Conduit (ARTICLE 350)

Flexible metal conduit is a listed metal raceway of circular cross section made of a helically wound, formed, interlocked metal strip. This wiring method is exposed or concealed. It is not typically permitted in wet locations, in hoistways, or in hazardous classified locations. It cannot be used in storage battery rooms, underground, embedded in poured concrete or aggregate, where subject to physical damage, or where exposed to materials having a deteriorating effect on the installed conductors.

Liquidtight Flexible Metal Conduit and Liquidtight Flexible Nonmetallic Conduit (ARTICLE 351)

Liquidtight flexible metal conduit is a listed raceway of circular cross section having an outer, liquidtight, nonmetallic, sunlight-resistant jacket over an inner flexible metal core.

Liquidtight flexible nonmetallic conduit is a listed raceway of circular cross section of various types:

- LFNC-A: A smooth seamless inner core and cover bonded together and having one or more reinforcement layers between the core and cover.

- LFNC-B: A smooth inner surface with integral reinforcement within the conduit wall.

- LFNC-C: A corrugated internal and external surface without integral reinforcement within the conduit wall.

This wiring method can be used where conditions of installation, operation, or maintenance require flexibility or protection from liquids, vapors, or solids. It cannot be used where subject to physical damage or where any combination of ambient and conductor temperature will produce an operating temperature in excess of the rating of the conduit.

Surface Metal Raceways (ARTICLE 352 A.)

A surface metal raceway is a metal raceway of various-shaped cross sections that is mounted on the surfaces of walls and ceilings. They have removable covers and the conductors are laid in these raceways after the raceway is assembled. This wiring method is only permitted in dry locations. This method cannot be used where subject to severe physical damage, for circuits of 300 V or more unless the metal is at least 0.04 inches thick, in corrosive environments, or in hoistways. It is usually not permitted to be used in concealed locations. This type of raceway is typically not used in hazardous (classified) locations except Class I, Division 2 (see Figure 6.7 for a classification description).

Surface Nonmetallic Raceways (ARTICLE 352 B.)

A surface nonmetallic raceway is the same as a surface metal raceway, but made of nonmetallic materials. It is not permitted where concealed, where subject to severe physical damage, for circuits of 300 V or more (unless specifically listed for higher voltages), or in hoistways. Nor is it permitted where subject to ambient temperatures exceeding those for which the nonmetallic raceway is listed. Nor is it to be used for conductors whose insulation temperature limitations would exceed those for which the nonmetallic raceway is listed. This type of raceway is typically not used in hazardous (classified) locations except Class I, Division 2 (see Figure 6.7 for a classification description).

Strut-Type Channel Raceway (ARTICLE 352 C.)

Strut-type channel raceways are formed of metal that are resistant to moisture or protected by corrosion protection. They are made of galvanized, stainless, enameled, or PVC-coated steel or of aluminum. Covers can be metallic or nonmetallic. They are used only where exposed, in dry locations, for circuits of 600 V or less, and in corrosive locations where protected by suitable finishes. They cannot be used where concealed.

Underfloor Raceways (ARTICLE 354)

Underfloor raceways are made of various materials and cross sections and are built into the floor. They may be permanently covered or may have removable covers. They are typically used in office space with movable partitions, exposition halls, and in retail stores. They are not typically used in hazardous (classified) locations and cannot be used in corrosive atmospheres.

Cellular Metal Floor Raceways (ARTICLE 356)

Cellular metal floor raceways are made of the hollow spaces of prefabricated metal floor cells for use in steel-frame buildings. This wiring method is typically not used in hazardous (classified) locations and cannot be used in corrosive environments.

Cellular Concrete Floor Raceways (ARTICLE 358)

Cellular concrete floor raceways are made of the hollow spaces in floors constructed of precast cellular concrete slabs, together with suitable metal fittings designed to provide access to the floor cells. This wiring method is typically not used in hazardous (classified) locations and cannot be used in corrosive environments.

Metal Wireways and Nonmetallic Wireways (ARTICLE 362)

Wireways are sheet metal troughs with hinged or removable covers for housing and protecting electric wires and cable and in which conductors are laid in place after the wireway has been installed as a complete system. They are similar to surface-mounted raceways. They can be used in exposed or concealed locations. They can be installed in Class I and II, Division 2 locations for intrinsically safe wiring. They cannot be used where subject to severe physical damage or corrosive vapor.

Busways (ARTICLE 364)

A busway is a grounded metal enclosure containing factory-mounted, bare or insulated conductors, which are usually copper or aluminum bars, rods, or tubes. They are used when located in the open and visible, or they can be installed behind access panels with certain provisions. They cannot be used where subject to severe physical damage, in corrosive environments, in hoistways, in hazardous (classified) locations unless listed for such use, in the outdoors, or in wet or damp locations unless listed for such use.

Cables

Installation of some cables is permitted in the open (i.e., outside of raceways or cable trays) as long as the cables are otherwise adequately supported. Armored Cable (Type AC) (Article 333),

Metal Clad Cable (Type MC) (Article 334), Non-metallic Sheathed Cable (Types NM, NMC, and NMS) (Article 336), and Mineral Insulated Metal Sheathed Cable (Type MI) (Article 330) are examples of cables that are permitted to be installed in the open as long as they are supported as described in the NEC.

OSHA (29 CFR 1910.305(g)) and the NEC (Article 400-7) limit the uses of flexible cords and cables to those instances where flexibility is needed for:

- Pendants
- Wiring of fixtures
- Connection of portable lamps or appliances
- Elevator cables
- Wiring of cranes and hoists
- Connection of stationary equipment to facilitate their frequent interchange
- Prevention of the transmission of noise or vibration
- Appliances where the fastening means and mechanical connections are designed to permit removal for maintenance and repair
- Data processing cables approved as a part of the data processing system

It should be noted that frequent interchange is undefined; however, there are OSHA interpretations limiting temporary wiring to at most 90 days. One rule of thumb is that interchange should be at least once per quarter in order to be considered frequent.

The NEC applies to the facility's wiring, not to the wiring of industrial machines. The Electrical Standard for Industrial Machinery (NFPA 79) applies to the wiring of machinery. Of particular importance to the safety professional is the machinery wiring installed at the facility. Such wiring may be between associated pieces of equipment in a machining system or in a custom-made control panel. For example, the wiring of conveyors and their sensors and controllers to other industrial machines is governed by NFPA 79. Whether employees or contractors perform this wiring, , the safety professional should ensure that the wiring methods are appropriate for the intended conditions. NFPA 79, paragraph 16.3.1 requires that conductors and their connections external to the control panel enclosure be totally enclosed in suitable raceways or enclosures, unless required for use with control pendants, for equipment movement, or to complete connections to normally stationary motors, limit switches, or other externally mounted devices.

Whether for facility wiring or machinery wiring, flexible wiring should be reserved for applications where flexibility is needed, and all wiring methods should conform to applicable standards. The safety professional should work with engineering and maintenance personnel to ensure adequate consideration and compliance with those standards.

Planning for Ease of De-energized Maintenance

When connecting machinery to electrical power, consider adequate electrical disconnects to facilitate lockout/tagout in accordance with the Control of Hazardous Energy (29 CFR 1910.147). There must be a readily accessible, easily identifiable, lockable electrical disconnect for each machine that can be worked on independently. When machining systems or processes have multiple pieces of equipment, determine whether each piece of equipment can be worked on while associated equipment remains running. The answer to this question is not always easy to identify.

Process engineering personnel responsible for the design and specification of the system typically view all equipment as critical to the design and often say that independent electrical disconnects in addition to the main electrical disconnect are not needed. Production management personnel often find ways to make acceptable product temporarily while individual pieces of equipment are not functioning, or else they decide to continue to make scrap for a short time while repairs are made to reduce total down time. Either way, production requirements often necessitate working on equipment while not locked out if adequate electrical disconnects have not been provided.

If a suitable solution can be identified, then a piece of equipment might be able to be worked on while other equipment remains in operation. However, if other equipment in the vicinity pose hazards to maintenance personnel, then the subject equipment cannot be worked on while the rest of the system remains operational. If a piece of equipment is determined to be independent, it needs to have a separate lockable electrical disconnect. Avoid placing this disconnect inside electrical cabinets that will have exposed live parts when the rest of the system is in operation. Place the lockable electrical disconnect as close as possible to the subject equipment, outside of electrical cabinets.

Elements of a Safe Electrical Design (What the Safety Professional Needs to Know)

The safety professional with responsibility to develop and oversee an Electrical Safety Program must know what questions to ask. It is unlikely that a safety professional for the facility will be an electrical engineer or an electrician. Therefore, the safety professional will not be able to determine the adequacy of hazard controls without technical assistance. However, he/she needs a basic knowledge of the elements of safe design so that he/she can ensure those elements are not overlooked or inadequately addressed by the technical staff.

If the NEC adopted by the local building code is being adhered to, compliance with OSHA Subpart S standards on electrical equipment will not be an issue. However, to help the safety professional correlate OSHA requirements with the NEC, Table 6.1 is provided. The articles in the table reflect the organization of the 2000 edition of the NEC.

Rating of Conductors

Conductors have current and voltage ratings to ensure that they are used only in applications for which they were designed. If an inappropriate conductor is used in a circuit with too high a voltage, the insulation of the conductor will break down. If too high a current is passed through a conductor, the conductor will overheat, damage the insulation, and possibly causing a fire. Ensure the technical staff have addressed adequate conductor size and rating for all expected loads as required by the NEC. This means that decisions made early in the project development must be revisited with every project change that can affect the required current of the system. See the discussion below on Safety Review Prior to Power-On.

Overcurrent Protection

Regardless of the appropriateness of the conductors, overcurrent protection is required for all feeders and branch circuits to prevent a fire in the case of an electrical fault. There are specific requirements in the NEC as to what rating the overcurrent protection must have for specific applications. Additionally, the equipment manufacturer usually specifies the overcurrent protection that must be provided for the equipment. Ensure that the technical staff has adequately addressed overcurrent protection in accordance with the NEC and the manufacturer's specifications.

Mechanical Support and Clearance

The NEC states specific requirements for physically supporting various wiring methods. It is essential that the chosen wiring method be implemented in compliance with the applicable NEC adopted by the local authority having jurisdiction. Ensure the NEC has been consulted for the wiring method chosen and that specific requirements for support have been met. This is one area where the safety professional can look up the requirements and visually verify proper implementation.

Electrical Integrity

The NEC and OSHA regulations require sound electrical connections at all points. Generally, electric terminals are provided on equipment for the purpose of connecting them to electric power. Wires are to be wrapped clockwise around terminal screws so that tightening will not disengage them. The wire is to go at least three-quarters of the way around the terminal screw. The wires are to be stripped of insulation at the point of connection, but should not be stripped of insulation so as to unduly expose the conductors. The connections should be tight so as to maintain continuity if wires are physically disturbed. Twisting connections are to use approved wire nuts inside approved junction boxes or electrical enclosures. Taping the wires together is not sufficient. Soldered connections must be reinsulated or located inside approved junction boxes or electrical enclosures.

Table 6.1: Correlation between OSHA Requirements and the NEC

Subpart S Paragraph	NEC Article
303	110
304(a)	200
304(b)	210
304(c)	225
304(d)	230
304(e)	240
304(f)	250
305(a)	300, 305, 318-365
305(b)	373
305(c)	380
305(d)	384
305(e)	Chapter 3, 4
305(f)	310
305(g)	400A
305(h)	400C
305(i)	402
305(j)	410, 422, 430, 450, 460, 480
306(a)	600
306(b)	610
306(c)	620
306(d)	630
306(e)	645
306(f)	660
306(g)	665
306(h)	668
306(i)	675
306(j)	680
307	Chapter 5
308(a)	110, 300B, 490
308(b)	700
308(c)	725
308(d)	760
308(e)	Chapter 8

Repairs may be made to cable sizes No.12 and larger as long as the insulation characteristics are preserved to original manufacturer's specifications.

Ground Fault Protection

A ground fault is a loss of the ground path. The loss of the ground path can result in electrocution when a person completes the circuit to ground. This is much easier to do in wet locations such as kitchens, bathrooms, and utility rooms. Therefore, protection from ground faults is required in wet and outdoor locations. Protection from a ground fault is provided by a Ground Fault Circuit Interrupter (GFCI). GFCIs sense the presence of a ground fault and open the circuit to prevent electrocution.

Grounding

The grounding requirements are many and varied in the NEC. However, in general all equipment enclosures and frames must be grounded. Grounding must be provided by the most direct path to ground practical. The size of the grounding conductor must be sufficient to carry fault current without overheating or raising the voltage to dangerous levels. Ensure the technical staff has addressed grounding as required by the NEC.

Segregation and Shielding

The NEC restricts the number, voltage, and types of conductors that can be co-located in cable runs or cable trays. The reason is to provide for adequate natural ventilation to keep the conductors from overheating. Ensure the technical staff has addressed the amount of cables in each cable run or tray. If it looks like the cable tray or run is filled with cables or there may not be adequate natural ventilation for each cable, question the adequacy of the installation. Look up the NEC requirements or ask the technical staff for verification that the installation is adequate.

Suitability for Location (Hazardous Locations, Wet Locations)

Electrical installations and equipment must be appropriate for the environmental conditions to which they are subjected. Of major concern in industrial plants are wet or damp locations and hazardous (classified) locations. Outdoors locations are always considered wet locations. Hazardous (classified) locations are areas where there is a potential for a flammable or combustible atmosphere to exist. Areas where flammable or combustible liquids or gases are stored, handled, piped, used, or disposed of are Class I hazardous locations. Areas where combustible dusts are stored, handled, or used are likely to be Class II hazardous locations. Areas where combustible, easily ignitable fibers or flyings are present, but in which such fibers or flyings are not likely to be in suspension in the air in quantities sufficient to produce ignitable mixtures are Class III hazardous locations. Figure 6.7 depicts the different classifications of hazardous locations. For more detailed information concerning the extent of hazardous locations for specific types of machines, materials, and industries, see Appendix B and the following standards:

- Flammable and Combustible Liquids Code (NFPA 30-1996)Standard for Drycleaning Plants (NFPA 32-1996)

- Standard for Spray Application Using Flammable or Combustible Materials (NFPA 33-1995)

- Standard for Dipping and Coating Processes Using Flammable or Combustible Liquids (NFPA 34-1995)

- Standard for the Manufacture of Organic Coatings (NFPA 35-1995)

- Standard for Solvent Extraction Plants (NFPA 36-1997)

- Standard on Fire Protection for Laboratories Using Chemicals (NFPA 45-1996)

- Standard for Gaseous Hydrogen Systems at Consumer Sites (NFPA 50A-1994)

- Standard for Liquefied Hydrogen Systems at Consumer Sites (NFPA 50B-1994)

- Liquefied Petroleum Gas Code (NFPA 58-1998)

- Standard for the Storage and Handling of Liquefied Petroleum Gases at Utility Gas Plants (NFPA 59-1998)

- Recommended Practice for the Classification of Flammable Liquids, Gases, or Vapors and of Hazardous (Classified) Locations for Electrical Installations in Chemical Process Areas (NFPA 497-1997)

- Recommended Practice for the Classification of Combustible Dusts and of Hazardous (Classified) Locations for Electrical Installations in Chemical Process Areas (NFPA 499-1997)

- Standard for Fire Protection in Wastewater Treatment and Collection Facilities (NFPA 820-1995)

- Recommended Practice for Classification of Locations for Electrical Installations at Petroleum Facilities Classified as Class I, Division 1 and Division 2 (ANSI/API RP500-1997)

- Area Classification In Hazardous (Classified) Dust Locations (ANSI/ISA-S12.10-1988)

There are significant differences between the classes of hazardous locations. Equipment suitable for Class I may not be suitable for Class II. The biggest reason for this is that dusts may build on the surface of equipment, thermally insulate that equipment, and heat up to its autoignition temperature. Vapors and gases do not so insulate equipment. Hazardous locations are also subdivided into Division 1 or 2 dependent upon the potential for a hazardous atmosphere to exist. Equipment rated for Division 2 locations only is not suitable for Division 1 locations. Division 1 equipment is expected to operate in a hazardous atmosphere. Division 2 equipment must prevent ignition only long enough for operators to take appropriate action to eliminate the hazardous atmosphere or shut down the electrical equipment. Finally, equipment may only be rated for a particular group(s) of substances within a class because different materials have different ignition temperatures, minimum ignition energies, and conductivities. It is essential that the right equipment be chosen for the materials present.

Class I: Flammable and combustible vapors or gases
Class I, Division 1: A location in which: • Ignitable concentrations of flammable gases or vapors can exist under normal operating conditions, or   • Ignitable concentrations of such gases or vapors may exist frequently because of repair, maintenance operations, or leakage, or   • Breakdown or faulty operation of equipment or processes might release ignitable concentrations of flammable gases or vapors, and might also cause simultaneous failure of electrical equipment in such a way as to directly cause the electrical equipment to become a source of ignition.
Class I. Division 2: A location in which: • Volatile flammable liquids or flammable gases are handled, processed, or used, but in which the liquids, vapors, or gases will normally be confined within closed containers or closed systems from which they can escape only in case of accidental rupture or breakdown of such containers or systems, or in case of abnormal operation of equipment, or   • Ignitable concentrations of gases or vapors are normally prevented by positive mechanical ventilation, and which might become hazardous through failure or abnormal operation of the ventilating equipment, or   • That is adjacent to a Class I, Division 1 location, and to which ignitable concentrations of gases or vapors might occasionally be communicated unless such communication is prevented by adequate positive-pressure ventilation from a source of clean air, and effective safeguards against ventilation failure are provided.
Class I, Group A: Acetylene
Class I, Group B: Vapors or gases with properties similar to hydrogen (maximum experimental safe gap (MESG) less than or equal to 0.45 mm, a minimum igniting current ratio (MIC ratio) less than or equal to 0.40)
Class I, Group C: Vapors or gases with properties similar to ethylene (MESG value greater than 0.45 mm and less than or equal to 0.75 mm, or a MIC ratio greater than 0.40 and less than or equal to 0.80)
Class I, Group D: Vapors or gases with properties similar to propane (MESG value greater than 0.75 mm or a MIC ratio greater than 0.80)
Class II: Combustible dusts
Class II, Division 1: A location in which: • Combustible dust is in the air under normal operating conditions in quantities sufficient to produce explosive or ignitable mixtures, or   • Mechanical failure or abnormal operation of machinery or equipment might cause such explosive or ignitable mixtures to be produced, and might also provide a source of ignition through simultaneous failure of electric equipment, operation of protection devices, or from other causes, or   • Combustible dusts of an electrically conductive nature may be present in hazardous quantities.

Figure 6.7: Classifications of Hazardous Locations

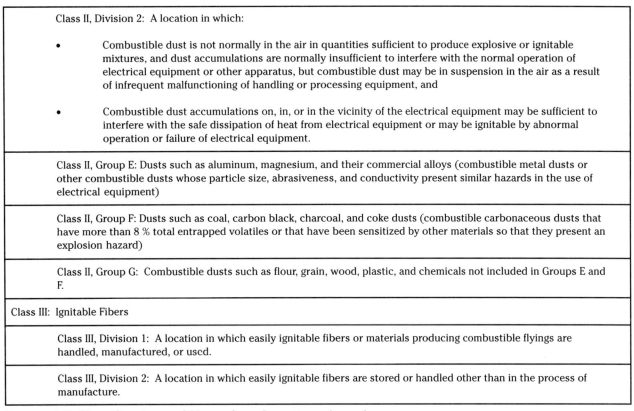

Class II, Division 2: A location in which:

- Combustible dust is not normally in the air in quantities sufficient to produce explosive or ignitable mixtures, and dust accumulations are normally insufficient to interfere with the normal operation of electrical equipment or other apparatus, but combustible dust may be in suspension in the air as a result of infrequent malfunctioning of handling or processing equipment, and

- Combustible dust accumulations on, in, or in the vicinity of the electrical equipment may be sufficient to interfere with the safe dissipation of heat from electrical equipment or may be ignitable by abnormal operation or failure of electrical equipment.

Class II, Group E: Dusts such as aluminum, magnesium, and their commercial alloys (combustible metal dusts or other combustible dusts whose particle size, abrasiveness, and conductivity present similar hazards in the use of electrical equipment)

Class II, Group F: Dusts such as coal, carbon black, charcoal, and coke dusts (combustible carbonaceous dusts that have more than 8 % total entrapped volatiles or that have been sensitized by other materials so that they present an explosion hazard)

Class II, Group G: Combustible dusts such as flour, grain, wood, plastic, and chemicals not included in Groups E and F.

Class III: Ignitable Fibers

Class III, Division 1: A location in which easily ignitable fibers or materials producing combustible flyings are handled, manufactured, or used.

Class III, Division 2: A location in which easily ignitable fibers are stored or handled other than in the process of manufacture.

Figure 6.7 Classifications of Hazardous Locations (cont.)

Ensure all electrical equipment and wiring methods chosen are suitable for their environment. Equipment will be labeled by the manufacturer as suitable for wet or damp locations, outdoor use, Class I, II, or III areas, Division 1 or 2 areas, and specific groups of materials A through G. Ensure that technical staff considers the adequacy of electrical installations when making changes to the process that could affect the class, division, or group of the surrounding area.

Chapter Seven

Periodic Electrical Safety Inspection and Preventive Maintenance

Overview

Another essential aspect of any electrical safety program is periodic inspection. The program should be reviewed as changes to electrical installations are made but also should be reviewed annually to ensure that significant changes have not been implemented without safety management review. Additionally, all levels of management should tour the electrical installations to identify any needed improvements or repairs. Compliance audit of electrical installations against the NEC is not practical. The review must incorporate observations that are easy to make; otherwise, management personnel will not support the process. Periodic inspection is part of the process of continuous improvement essential to a good electrical safety program.

Safety Inspections

The responsibility for safe electrical systems in the workplace is ultimately the plant manager's. This responsibility should be delegated to the department managers, shift coordinators, and cell leaders. The front line supervisor should inspect his area of responsibility at least daily. His/her superiors should periodically inspect all areas under their responsibility. The safety professional and safety team should incorporate looking for electrical hazards into their inspections. And the plant manager should get out into the plant to inspect the safety of the workplace. Whenever a hazard is identified, it should be reported to the proper authority for correction. These inspections serve a two-fold purpose: to identify electrical hazards in the workplace, and to verify that the electrical safety program is sufficient to control electrical hazards. Where an electrical hazard is identified, it should be investigated to determine how the condition developed and how it can be prevented from developing again. The checklist in Figure 7.1 is provided as an example of the kinds of items that can be directly and easily observed throughout the facility. This checklist should be used by safety professionals, supervisors, management, and safety team members while inspecting areas and equipment for electrical safety.

It is essential that deficiencies be corrected in a timely manner. There must be an administrative system that will assign a responsible person to the corrective action, schedule a reasonable completion date, and track the status of the action items. This administrative system must efficiently communicate action items to the responsible people and, in the case of high-risk issues, to the appropriate level of management. Additionally, the system must communicate the status of action items to the appropriate level of management periodically and on demand. A simple spreadsheet or table can be used to document the action items. However, there is probably already a system in place to accomplish these tasks. The question is how well does that system perform, and can it be improved. Who is responsible to input the data? How often is a report of open items generated? Who receives the status report? Are there any performance objectives in place for personnel to accomplish these tasks? There is probably a work authorization or work request procedure in place at the facility. However, this system is usually dedicated to maintenance activities. Not all deficiencies with a safety impact will be the responsibility of the maintenance department to correct. The current system must be expanded to incorporate corrective actions for all types of deficiencies, or a separate system must be created to track other types of action items. These issues must be addressed to ensure that safety issues, once identified, do not fall through the cracks.

Preventive Maintenance

Preventive maintenance (PM) is the element of the electrical safety program that identifies potential failure of hazard controls and corrects them before a hazardous situation develops. PM can reduce the likelihood of electrical failures leading to electrocution and fire. PM is essential for systems requiring active ventilation for thermal control and where electrical installations are located in hazardous (classified) locations. Such locations have or could have combustible mixtures of flammable or combustible materials and air. Electrical equipment and facility wiring should be incorporated into the facility's PM program.

PM includes cleaning, inspecting, testing, and lubricating. A good PM program consists of:

- Identifying electrical equipment that requires or would benefit from PM
- Identifying what PM is needed
- Developing procedures to accomplish the PM
- Training maintenance personnel how to perform the PM
- Scheduling the PM
- Tracking performance of the PM

Motors, transformers, disconnects, overcurrent protection devices, interlocks, electrical conductors, grounding conductors, motor control centers, flexible cables, fittings, and enclosures are examples of equipment that should be periodically inspected. Electrical enclosures that require fans to keep them cool should be placed in the PM program. Areas of critical importance are hazardous classified areas that contain or could contain combustible atmospheres.

Electrical equipment and facility wiring should be incorporated into the facility's preventive maintenance program. The Recommended Practice for Electrical Equipment Maintenance (NFPA 70B) and Part III of the Standard for Electrical Safety Requirements for Employee Workplaces (NFPA 70E) provide guidance for the development of a PM program for electrical equipment.

❏ Electrical equipment is marked with the manufacturer's name, trademark, or other descriptive marking.

❏ Electrical equipment is marked with voltage, current, wattage, or other appropriate ratings.

❏ Equipment that is wet with water, oil, or other liquid is rated for a wet location.

❏ Equipment is protected by GFCI if it is located in bathrooms, on rooftops, in wet locations, outdoors, or within 6 feet of sources of water.

❏ Equipment is rated for outdoor use if it exposed to the outdoors.

❏ Equipment does not have accumulations of dust, chips, or other solids on it or under it.

❏ Hot equipment does not have combustible material (paper, wood, or plastic) stored immediately adjacent to it.

❏ Equipment in flammable or combustible liquid or gas storage areas or oxygen storage areas is rated Class I, Division 2.

❏ Hot or arcing equipment is not located within 20 feet of stored oxygen or flammable gas cylinders.

❏ Equipment is rated for appropriate hazardous location where flammable or combustible liquids or combustible dusts are used, created in a process, dispensed or otherwise opened to atmosphere, or stored in large quantities.

❏ Equipment including raceways is rated for a corrosive environment where it is exposed to acids or caustic mists, splashes, or drips from chemical piping/tubing.

❏ Electrical equipment has a readily identifiable disconnecting means. Walk up to a machine or motor and find its disconnect.

❏ Electrical disconnects are marked as to their purpose, or else the purpose is evident.

❏ Circuit breakers and fuses inside panels and the panels themselves are labeled as to their purpose.

❏ Circuit breakers are labeled as to what they power.

❏ All electrical panel doors can be opened 90° without moving other equipment or storage out of the way.

❏ The space in front of panelboards, switchboards, and switchgear is at least 30 inches wide or the entire width of the panel, whichever is larger.

❏ The space in front of panelboards, switchboards, and switchgear is at least 6 feet 6 inches tall.

❏ The space in front of panelboards, switchboards, and switchgear is at least as deep as the tables below indicate::

Clearance in Front of Electrical Enclosures Opening toward Insulated Parts or Materials

50 – 600 V	At least 3 feet

Clearance in Front of Electrical Enclosures Opening toward Grounded Parts

50 – 150 V	At least 3 feet
151 – 600 V	At least 3.5 feet

Clearance in Front of Electrical Enclosures Opening toward Exposed Live Parts

50 – 150 V	At least 3 feet
151 – 600 V	At least 3.5 feet

Figure 7.1: Periodic Electrical Safety Inspection Checklist

❏ There are no exposed live parts greater than 50 volts within reach from any floor or work platform (i.e., parts not more than 600 V are more than 8 feet above platform or more than 4 feet beyond platform guardrail).

❏ The covers are properly fastened on electrical boxes (no missing bolts or broken latches).

❏ Electrical boxes are located so that they will not be damaged (not in the aisle with motorized traffic).

❏ Electrical boxes are firmly fastened in place.

❏ Electrical boxes do not have holes or gaps in them larger than ¼ inch.

❏ Electrical boxes are not bent, dented, or crushed so that there is a danger of shorting components inside.

❏ Metal enclosures, raceways, and machine casings are grounded to earth ground by means of a continuous conducting path.

❏ Fixed machines are grounded directly to earth or by grounding conductor of sufficient size to earth.

Rating or Setting of Automatic Overcurrent Device in Circuit Ahead of Equipment, Conduit, etc., Not Exceeding (Amperes)	Size (AWG or kcmil) Copper
15	14
20	12
30 - 60	10

❏ Grounding path from one machine does not pass through another machine before reaching earth.

❏ Grounding path from one machine, equipment, of flammable storage cabinet does not pass through a flammable storage cabinet before reaching earth.

❏ Equipment grounding cables are firmly attached to an **appropriate** grounding electrode.

❏ Equipment with a three-prong plug is connected to power without interrupting the ground path.

❏ Conductors entering cabinets, boxes, and enclosures are protected from abrasion and being cut.

❏ Disconnects are oriented so that On is up and OFF is down.

❏ Flexible cords and cables are only used where flexibility is required such as for pendants, fixture wiring, portable lamps or appliances, elevator cables, cranes and hoists, frequently interchanged stationary equipment, noise and vibration isolation, appliances that are removed for maintenance and repair, and data processing cables.

❏ Flexible cords and cables have strain relief where attached to devices and fittings.

❏ Flexible cords do not have splices or unapproved repairs.

❏ Flexible cords are not attached or supported on fixed structures.

❏ Flexible cords are in good condition.

❏ Electric motors are clean.

❏ Ventilation slots or holes for electrical equipment are not obstructed.

❏ Ventilation filters for electrical equipment are clean.

❏ There are no locations where wires are not protected by rigid or flexible conduit, enclosure, cable tray, or flexible cable jacket.

❏ Conduit is supported horizontally and vertically.

❏ Cable is supported by cable tray.

Figure 7.1: Periodic Electrical Safety Inspection Checklist (cont.)

Electricity Basics and Electrical Hazards

The following discussions are purposefully brief. For more information, see any good physics or engineering text on electricity.

Fundamental Characteristics of Electricity

Charge is a fundamental characteristic of the atomic particles that make up matter. It can be positive or negative. Electrons have negative charge. Holes have positive charge. Holes are spaces in the molecular structure that are normally filled with electrons, but have had the electrons stripped away. Thus, the molecular structure is positively charged. Static electricity is built up as a charge, usually through friction between two dissimilar materials (e.g., plastic and air, flammable liquids and air, leather shoes and carpet, or car upholstery and clothing). Like charges repel one another, and opposite charges attract. The unit of measure of charge is the Coulomb.

Current is the movement of positive or negative charges. Current is caused by a difference in electrical potential (voltage). The voltage tries to combine negative and positive charges to reduce the net charge. It is current that causes shock. Charges moving through the body create heat and subsequent burns. The current also disrupts the neurological functions that are essentially electrical in nature. The unit of measure of current is the ampere or amp (A). An amp is 1 Coulomb of charge per second passing a given point.

Voltage is a measure of electrical potential. It is analogous to pressure when considering fluid flow. Fluids move from high pressure to low pressure. Positive charges move from high positive voltage to low positive or negative voltage. The unit of measure of voltage is the volt (V).

Circuits and Current Flow (AC & DC)

A circuit is a path along which current can flow. The flow is driven by voltage. Wires, bus bars, and electrodes are conductors. Circuits are often thought of as linear from high voltage to low voltage, but circuits are loops. Chemical work (i.e., batteries) or mechanical work (i.e., generators) create electrical potential. The generator or battery raises charges from zero potential (i.e., ground) to higher potential. If there is a circuit path to a lower potential, current will flow. If there is no circuit path to a lower potential, the circuit is said to be "open." In an open circuit, charges build up as on a capacitor or as in the case of static electricity. The buildup of charge waits for any path to lower electrical potential to present itself. When the charges finally reach the place where they came from, the circuit is complete. Charges cannot be removed from one place and deposited in another indefinitely without causing a buildup of charge. (See Grounding below.) If the high-voltage side of the circuit is connected directly to the low-voltage side, the circuit is said to be "shorted." In a short circuit, current readily flows because there are no loads to resist the flow. Very high currents can be achieved, causing damage to the equipment and fires.

There are two basic types of circuits: alternating current (AC) and direct current (DC). In DC circuits, the battery or DC generator always creates a potential such that one side of the circuit is always at a higher potential than the other. Current always flows in the same direction. In AC circuits, the AC generator alternates the voltage between positive and negative. Each side of the circuit takes turns being at higher potential than the other. Current therefore alternates, first flowing in one direction, then the other. The frequency at which the current changes directions is measured in Hertz (Hz). Electrical power in the United States is generated and transmitted at 60 Hz. In Europe, it is common for AC power to be supplied at 55 Hz.

Grounding

Grounding connects a conductor to a large conductive chassis (e.g., a metal frame) or to the earth. Circuits are connected to earth ground to provide a large enough source or "sink" so that charges will not buildup appreciably due to the current flow. Two types of grounding are used. The grounded conductor is the neutral wire that completes the circuit from the load to ground. (i.e., charge is raised to high potential by the power source and flows to the load through the primary conductor. Current continues through the grounded conductor and returns to earth or chassis ground.) The second type of ground is a grounding conductor. This conductor is the third wire in a three-wire outlet circuit common in homes and industry. The grounding conductor is not intended to carry current under normal circumstances. The grounding conductor will only carry current if the power is inadvertently shorted to the equipment case, enclosure, or frame. The grounding conductor provides a direct path back to ground so that anyone touching the equipment case, enclosure, or frame will not be injured. The grounding conductor is a safety feature and, as such, must be carefully preserved.

It is essential that the grounding path not be interrupted. Using three-wire to two-wire adapters without connecting the grounding tab to ground is prohibited. Removing the grounding prong on a three-wire plug is prohibited. Bending the grounding prong out of the way to plug it into a two-wire extension cord or outlet is prohibited. Failing to connect equipment housings, electrical enclosures and panels, and equipment frames to ground is also prohibited. See the NEC for the requirements on what equipment must be grounded.

Resistance and Heat

The flow of current through conductors, appliances, and equipment generates heat because these circuit components have resistance. They resist the flow of current. Only a superconductor has no resistance to current flow. Thus, equipment and conductors get hot when they are powered. If the equipment or conductors get too hot, they will be damaged or may start a fire. Mechanical equipment also generates heat because of friction between moving parts. In general, the hotter the conductor or equipment, the more resistance it has. Therefore, the heat generated by current-carrying components must be removed. Natural ventilation or forced ventilation can be used to remove heat. Some equipment is cooled with refrigerated air or water to remove its heat. When too many heat-producing components or conductors are present in a space, it can easily generate more heat than can be removed. Therefore, there are limits as to how many conductors are allowed to be in a cable tray or run through a hole in a cable trace. When equipment is allowed to become dirty or material is allowed to be piled up around equipment, natural ventilation becomes ineffective. Therefore, equipment must be kept clean and free from debris. If too much current is passed through a conductor, it will overheat. For this reason conductors have current ratings or limitations that must be observed. And that is why plugging in two many loads into a receptacle outlet causes a fire.

Impedance (Capacitance and Inductance)

Capacitance is the ability of a component to store electric charge. Capacitors are designed for this purpose. Capacitance is measured in units of farads (f). The larger the capacitance, the more charge the capacitor will hold. Capacitors will hold their charge until a current path is provided or until the charge dissipates into the surrounding environment. Some capacitors are designed with a current path that quickly discharges it when the power is OFF. Other capacitors can hold their charge for minutes or even hours after power is turned OFF. These capacitors must be manually discharged by qualified employees prior to working on the circuit.

Inductance is the ability of a component to store energy from current in magnetic fields. Inductance is measured in units of Henrys (H). Electromagnetic coils, large motors, and solenoids have significant inductance. Inductors will hold their magnetic field until a current path is provided or until the magnetic field dissipates into the surrounding environment. Some inductive components are designed with a current path that quickly de-energizes it when the power is OFF. Other inductive components can hold a magnetic field for many minutes after the power is turned OFF. When a current path is provided, the magnetic field will collapse inducing a current to flow. Therefore, inductive components can be as dangerous as capacitors.

Capacitance, inductance, and resistance are forms of impedance. In AC circuits, capacitance and inductance are out of phase with each other. Capacitance is used to counteract the inductance of large motors so that the overall impedance of the circuit is low.

Arc Flash

Arc flash is a rapidly expanding ball of gas and fume caused when electrical conductors are vaporized by fault current.

Electrical Shock

Electric shock is the most common hazard associated with electricity. Shock is caused by current passing through the body. The current causes burns along the path and disrupts neurological activity. Electric shock can cause muscle spasms that are so strong they can tear the muscle apart, stop the heart, or stop breathing. Electric shock can also stop brain function. It is current that causes the damage not voltage. It takes some minimum voltage to drive current through the skin. Wet skin has less resistance than dry skin. Generally 50 V is considered low voltage insufficient to cause shock. However, at sufficient voltage to penetrate the skin (as little as 30 V for wet skin or less for damaged skin), a current as small as 18 mA across the chest can cause respiratory arrest. A current as little as 70 mA can cause heart can cause ventricular fibrillation where the heart has no effective heartbeat.

Static Electric and Magnetic Fields

Static fields are fields that are not changing. They are constant. Static electric fields (E), if strong enough, can strip electrons away from atoms and molecules and propel the atomic nuclei in the opposite direction. Therefore, these fields can cause damage to cells. This technique is used to separate hydrogen and oxygen from water. Capacitors have a static electric field when they are not being charged or discharged. Batteries have static electric fields between their plates and poles. The strength of the E field is measured using an electric field survey meter that measures in volts/meter (V/m). The American Conference of Governmental Industrial Hygienists (ACGIH®) provides threshold limit values (TLVs®) and ceiling limits for static electric and magnetic fields and for non-ionizing radiation in their booklet *Threshold Limit Values for Chemical Substances and Physical Agents and Biological Exposure Indices*, available from ACGIH at 1330 Kemper Meadow Drive, Cincinnati, OH 45240-1634 or www.acgih.org.

Static and Sub-Radiofrequency Electric Fields	
0 – 100 Hz	25 kV/m ceiling

Static magnetic fields (B or H) can do much the same thing. However, certain materials are more sensitive to magnetic fields than others – namely iron and steel. Permanent magnets are used in industry to ensure metal fragments from machines do not get packaged with food products. Electromagnets are used to handle large pieces of iron and steel scrap, hold steel work-pieces in place during surface grinding, and many other uses. One of the largest dangers from static magnetic fields is that medical implants will be displaced within the body of the wearer. For that reason, the limits of exposure to static magnetic fields is much lower for this specific group of people. The strength of magnetic fields is measured using a Gauss meter that measures in Tesla (T) or milli-Tesla (mT). The ACGIH TLVs and ceiling limits for static magnetic fields are depicted in Table A.1.

Table A.1: TLVs and Ceiling Limits for Static Magnetic Fields

Static Magnetic Fields	8-hour TWA TLV	Ceiling Limit
Whole Body	60 mT	2 T
Limbs	600 mT	5 T
Medical electronic device wearers	—	0.5 mT

Non-Ionizing Radiation

AC circuits cause alternating electromagnetic fields to emanate from the circuit in all directions. These fields can be strong enough to induce current in other nearby circuits. This is called electromagnetic interference. Radio antennas, radar, and microwave equipment generate electromagnetic fields to perform their function. These emanating electromagnetic fields are non-ionizing radiation in the radiofrequency (RF) and sub-RF range. Exposure to non-ionizing radiation above recommended limits can cause burns to the skin, tissue, and organs it encounters or can induce currents in the body resulting in electric shock. Sunburn is a good example of non-ionizing radiation damage from the electromagnetic field in the ultraviolet range from the sun.

The strength of the electromagnetic energy flux density is measured using an electromagnetic radiation survey meter in milli-watts per square centimeter (mW/cm^2). Touch potentials are measured by special contact ammeters in milliamps (mA). Sources of non-ionizing radiation include RF welders and dryers, induction and high frequency (HF) welders, microwave ovens and dryers, dielectric dryers, plasma etch equipment, RF and microwave communication equipment, and radar.

OSHA sets the limit for electromagnetic energy flux density in its standard, Nonionizing Radiation (29 CFR 1910.97). For frequencies of electromagnetic energy between 10 mega-Hertz (MHz) and 100 giga-Hertz (GHz), the maximum power flux density is 10 mW/cm^2 to the whole body or to any part of the body for 0.1 hours or longer. This standard is based on old data and needs to be updated.

The ACGIH provides TLVs and ceiling limits for non-ionizing radiation. These TLVs are more stringent than OSHA for frequencies below 3 GHz. The current recommendations of the ACGIH should be ascertained by reviewing the section on Non-Ionizing Radiation published in the most recent edition of the *Threshold Limit Values for Chemical Substances and Physical Agents and Biological Exposure Indices*. Tables A.2 through A.5 are derived from information in the above-listed source.

Table A.2: Sub-Radiofrequency Magnetic Fields
(f = frequency in Hz)

1 – 300 Hz	Whole Body	60/f mT ceiling
	Arms and legs	300/f mT ceiling
	Hands and feet	600/f mT ceiling
300 – 30 kHz	—	0.2 mT ceiling

Table A.3: Sub-Radiofrequency Electric Fields
(f = frequency in Hz)

0 – 100 Hz	25 kV/m ceiling
100 – 4000 Hz	$2.5 \times 10^6/f$ V/m ceiling
4 – 30 kHz	625 V/m ceiling

Table A.4: Radiofrequency and Microwave TLVs Part A – Electromagnetic Fields
(f = frequency in MHz)

Frequency	Power Density, S (mW/cm²)	Electric Field Strength, E (V/m)	Magnetic Field Strength, H (A/m)	Averaging Time (minutes)
30 – 100 kHz	—	614	163	6
0.1 – 3 MHz	—	614	16.3/f	6
3 – 30 MHz	—	1842/f	16.3/f	6
30 – 100 MHz	—	61.4	16.3/f	6
100 – 300 MHz	1	61.4	0.163	6
0.3 – 3 GHz	f/300	—	—	6
3 – 15 GHz	10	—	—	6
15 – 300 GHz	10	—	—	616,000/f

Table A.5: Radiofrequency and Microwave TLVs Part B – Induced and Contact
Radiofrequency Currents Maximum Current (mA) (f = frequency in MHz)

Frequency	Through Both Feet	Through Either Foot	Contact	Averaging Time
30 – 100 kHz	2000 f	1000 f	1000 f	1 second
0.1 – 100 MHz	200	100	100	6 minutes

Hazardous (Classified) Locations

Hazardous (classified) locations are areas where a potential to have a flammable or combustible atmosphere is present. Such areas are highly susceptible to ignition sources such as electric sparks, arcs, and heat. It is essential that electrical installations in such areas be rated for those areas and the specific group of flammable or combustible substances used in those areas. See the Article 500 of the NEC for the requirements for electrical installations in such area. See the *Flammable and Combustible Liquid Code®* (NFPA 30) for descriptions of the extent of hazardous (classified) locations surrounding sources of flammable and combustible liquids. See the *Recommended Practice for the Classification of Combustible Dusts and of Hazardous (Classified) Locations for Electrical Installations in Chemical Process Areas* (NFPA 499 - 1997 ed.) for descriptions of the extent of hazardous (classified) locations surrounding sources of combustible dust.

See Figure 6.7 for a description of the classifications of hazardous locations.

For flammable and combustible liquids, the extent of a Class I area depends on the potential for a flammable vapor-air mixture to exist outside of process equipment or tanks. Factors that affect how far the Class I area extends are whether the flammable vapor-air mixture exists under normal conditions within the process equipment or tank, the pressure of the vapors or liquids within the process equipment or tank, and the ventilation effectiveness. All of the examples given below assume that there is sufficient ventilation to prevent a flammable vapors from concentrating to greater than 25% of their lower flammable limit (LFL). Even though ventilation is assumed to be adequate, the Class I area can be quite extensive.

Example 1

For equipment located indoors that can contain a flammable vapor-air mixture under normal conditions, the extent of the Division 1 area is out to 5 feet from any opening. The edge of the 5-foot boundary extends vertically downward to the floor because vapors of these liquids are heavier than air. The Division 2 area extends from 5 feet to 8 feet from such openings and for up to 3 feet above the floor or grade out to 25 feet. Any area below grade within the Division 2 area is considered Division 1.

For the same equipment located outdoors, the extent of the Division 1 area is out to 3 feet from any openings, and the Division 2 area is out to 8 feet from any openings and up to 3 feet above grade out to 10 feet. The reason for the reduction is that more effective ventilation is assumed outdoors.

Example 2

For filling drums and containers indoors or outdoors, the extent of the Division 1 area is out to 3 feet from the edge of the drum or container. The Division 2 area extends to 5 feet from the edge of the drum or container and up to 18 inches above the floor or grade for out to 10 feet. The reason for the apparent reduction in the extent of the area is because the flammable vapor-air mixture is not assured as in the case of equipment that is expected to have such a mixture present normally.

Example 3

For pump seals, bleeders, and withdrawal fittings on process equipment located indoors, the extent of the Division 2 area is out to 5 feet and up to 3 feet above the floor out to 25 feet. The reason for the increase in the extent of the Class I area beyond container-filling operations is that this equipment contains liquid under pressure.

For the same equipment located outdoors, the Division 2 area extends to only 3 feet and up to 18 inches above grade level out to 10 feet. Again the ventilation is assumed to be more effective outdoors.

The extent of Class II areas depends on similar factors to those discussed for Class I areas. Additionally, the following factors must be considered:

- Combustible material involved
- Bulk density of the material
- Particle sizes of the material
- Particle density
- Size of the leak opening
- Quantity of the release
- Dust removal system
- Housekeeping
- Presence of any hybrid mixture

Example 4

For open or semi-enclosed equipment processing or handling Group F or G dust indoors in an unrestricted or open area (i.e., no walls or other equipment nearby that could change or inhibit the ventilation), the extent of the Class II, Division 1 area is out to 20 feet in all directions from any opening. The Division 2 area extends beyond the Division 1 area out another 10 feet (i.e., 30 feet from any opening).

Types of Electrical Installation

Once a hazardous (classification) location is identified and its extent determined, the NEC, Chapter 5, establishes the requirements for the type of electrical installation that is suitable for the area. No matter what method is chosen, the circuit components and equipment must be marked as appropriate for the Class, Division, and Group of material in the area. Among the types of installation methods, circuits, and components that are used in such locations are the following:

Explosionproof Apparatus

Explosionproof apparatus will contain any explosion within the electrical enclosure, raceway, or equipment – preventing it from propagating outside of such equipment. It can be used for the Class I, Division 1 and 2 locations for which it is approved.

Dust Ignitionproof

Dust Ignitionproof apparatus will not get hot enough to ignite dust that accumulates on the outside of the equipment, nor will it permit dust into the equipment. It can be used for the Class II, Division 1 and 2 locations for which it is approved.

Dusttight

Dusttight apparatus will keep the dust out of the equipment. It can be used for the Class II, Division 2 and Class III locations for which it is approved.

Purged and Pressurized

Purged and pressurized enclosures and raceways are maintained under positive pressure so that flammable and combustible materials cannot enter. The NEC references another standard that describes how such a system is to be designed, installed, and operated—Standard for Purged and Pressurized Enclosures for Electrical Equipment (NFPA 496). It can be used for any hazardous (classified) location for which it is approved.

Intrinsically Safe Systems

Intrinsically safe apparatus and wiring is not capable of igniting the subject flammable or combustible material because the circuit has insufficient energy to do so. It can be used for in any hazardous (classified) location for which it is approved.

Nonincendive

Nonincendive apparatus, under normal operating conditions, is not capable of igniting the subject flammable or combustible material because its arcing and thermal effects are insufficient. It can be used for the Class I, Division 2; Class II, Division 2; and Class III locations for which it is approved.

Oil Immersion

Oil immersion components are literally immersed in oil to prevent atmosphere from contacting arcing or conductive parts. It can be used for current-interrupting contacts in Class I, Division 2 locations.

Hermetically Sealed

A hermetically sealed device is sealed against the entrance of an external atmosphere by soldering, brazing, welding, or the fusion of glass to metal. It can be used for current-interrupting contacts in Class I, Division 2 locations.

Standards for Electrical Safety

The following OSHA standards can be viewed and downloaded on the internet for free at www.osha.gov.

Electrical Protective Devices, Title 29, *Code of Federal Regulations*, Part 1910.137

Control of Hazardous Energy (Lockout/Tagout), Title 29, *Code of Federal Regulations*, Part 1910.147

Electrical, Title 29, *Code of Federal Regulations*, Part 1910, Subpart S

Electric Power Generation, Transmission, and Distribution, Title 29, *Code of Federal Regulations*, Part 1910.269

The following NFPA standards may be purchased individually or as a set from the National Fire Protection Association, 1 Batterymarch Park, P.O. Box 9101, Quincy, MA 02269-9101 or on the internet at www.nfpa.org. (NFPA standards that have been adopted by the local municipality or county are usually available for viewing at the government office with responsibility for enforcing such laws. Call the local building inspector or fire marshal to determine what codes are enforced.)

National Electrical Code® (NFPA 70-1999)

Flammable and Combustible Liquids Code® (NFPA 30-2000)

Recommended Practice for Electrical Equipment Maintenance (NFPA 70B-1998)

Standard for Electrical Safety Requirements for Employee Workplaces (NFPA 70E-2000)

Electrical Standard for Industrial Machinery (NFPA 79-1997)

Life Safety Code® (NFPA 101-2000)

Standard for Drycleaning Plants (NFPA 32-1996)

Standard for Spray Application Using Flammable or Combustible Materials (NFPA 33-1995)

Standard for Dipping and Coating Processes Using Flammable or Combustible Liquids (NFPA 34-1995)

Standard for the Manufacture of Organic Coatings (NFPA 35-1995)

Standard for Solvent Extraction Plants (NFPA 36-1997)

Standard on Fire Protection for Laboratories Using Chemicals (NFPA 45-1996)

Standard for Gaseous Hydrogen Systems at Consumer Sites (NFPA 50A-1994)

Standard for Liquefied Hydrogen Systems at Consumer Sites (NFPA 50B-1994)

Liquefied Petroleum Gas Code (NFPA 58-1998)

Standard for the Storage and Handling of Liquefied Petroleum Gases at Utility Gas Plants (NFPA 59-1998)

Standard for Purged and Pressurized Enclosures for Electrical Equipment (NFPA 496-1998)

Recommended Practice for the Classification of Flammable Liquids, Gases, or Vapors and of Hazardous (Classified) Locations for Electrical Installations in Chemical Process Areas (NFPA 497-1997)

Recommended Practice for the Classification of Combustible Dusts and of Hazardous (Classified) Locations for Electrical Installations in Chemical Process Areas (NFPA 499-1997)

Standard for Fire Protection in Wastewater Treatment and Collection Facilities (NFPA 820-1995)

The following ANSI standards may be purchased from Global Engineering Documents, 15 Inverness Way East, Englewood, CO 80112, (800) 854-7179, global.his.com or downloaded from the internet at www.ansi.org:

Recommended Practice for Classification of Locations for Electrical Installations at Petroleum Facilities Classified as Class I, Division 1 and Division 2 (ANSI/API RP500-1997)

Area Classification in Hazardous (Classified) Dust Locations (ANSI/ISA-S12.10-1988)

National Electrical Safety Code® (NESC) (ANSI C2), American National Standards Institute, 1819 L Street, NW, 6th Floor, Washington, DC, 20036

Safety Requirements for Portable Wood Ladders (ANSI A14.1), 1994

Safety Requirements for Fixed Ladders (ANSI A14.3), 1984

Safety Requirements for Job-Made Ladders (ANSI A14.4),1992

Safety Requirement for Portable Reinforced Plastic Ladders (ANSI A14.5), 1992

Series of Standards for Safety Signs and Tags (ANSI Z535), 1998

The following ASTM standards are published under ANSI and can be purchased as any other ANSI standard or directly from the American Society for Testing and Materials, 100 Barr Harbor Drive, West Conshohocken, PA 19428-2959, www.astm.org. Because federal OSHA adopted these ASTM standards by reference, they can be viewed at the Office of the Federal Register, 800 North Capitol Street, NW., Suite 700, Washington, DC, or at OSHA in Washington DC or at any OSHA regional office.

Standard Specification for Rubber Insulating Blankets (ASTM D 1048), American Society of Testing and Materials, 100 Barr Harbor Drive, West Conshohocken, PA 19428-2959, 1998

Standard Specification for Rubber Covers (ASTM D 1049), 1998

Standard Specification for Rubber Insulating Line Hoses (ASTM D 1050), 1990

Standard Specification for In-Service Care of Insulating Line Hose and Covers (ASTM F 478), 1992

Standard Specification for In-Service Care of Insulating Blankets (ASTM F 479), 1995

Standard Specification for Fiberglass-Reinforced Plastic (FRP) Rod and Tube Used; in Line Tools (ASTM F 711), 1989 (R 1997)

Standard Test Methods for Electrically Insulating Plastic Guard Equipment for Protection of Workers (ASTM F 712), 1988 (R 1995)

Standard Specification for Temporary Protective Grounds to Be Used on De-energized Electric Power Lines and Equipment (ASTM F 855), 1997

Standard Specification for Insulated and Insulating Hand Tools (ASTM F 1505), 1994

Suggested Readings:

Injury Facts®, 1999, National Safety Council, 1121 Spring Lake Drive, Itasca, IL 60143-3201, (630) 285-1121, www.nsc.org.

Jeffrey E. LaBelle, "What Do Accidents Truly Cost?,", Professional Safety, American Society of Safety Engineers, April 2000, 38-42. This article can be viewed online for free at www.asse.org under Professional Safety, Past Articles.

2000 TLVs and BEIs, American Conference of Governmental Industrial Hygienists, 1330 Kemper Meadow Drive, Cincinnati, OH 45240-1634, (513) 742-2020, www.acgih.org.

Recommended Practice for Electrical Power Distribution for Industrial Plants (IEEE 141), Institute of Electrical and Electronics Engineers, 445 Hoes Lane, Piscataway, NJ 08854, (800) 701-4333, www.ieee.org.

ACGIH. See American Conference of Governmental Industrial Hygienists.

Affected employee. An employee whose job requires operation of machinery being repaired or serviced under the requirements of this procedure (energy locked out/tagged out) or an employee who is required to work in an area where machinery is being serviced or repaired.

AIT. See Autoignition Temperature

Air Plenum. A compartment or chamber to which one or more air ducts are connected and that forms part of the air distribution system.

American Conference of Governmental Industrial Hygienists (ACGIH). An association that publishes exposure limits which are sometimes adopted in federal, state, and local regulations.

Appliances. Utilization equipment, generally other than industrial, normally built in standardized sizes or types, which is installed or connected as a unit to perform one or more functions, such as clothes washing, air conditioning, food mixing, or deep frying.

Authorized employee. An employee trained in the requirements of the regulations, standards, and company procedures concerning control of hazardous energy. This employee should be designated as an authorized employee by the plant manager or his authorized representative.

Autoignition Temperature (AIT). The minimum temperature required to initiate or cause self-sustained combustion of a solid, liquid, or gas independently of the heating or heated element.

Branch circuit. The circuit conductors between the final overcurrent device protecting the circuit and the outlet(s).

Busbar. Conductive metal bars, typically made of copper, used to distribute power in premises wiring systems inside a busway (see definition).

Busway. An approved assembly of conductors (i.e., busbars) in a completely enclosed, ventilated, protective metal housing. It is a type of electrical raceway (see definition).

Cabinet. An enclosure designed either for surface or flush mounting, and provided with a frame, mat, or trim in which a swinging door or doors are or may be hung.

Cable tray system. A cable tray system is a unit or assembly of units or sections, and associated fittings, made of metal or other noncombustible materials forming a rigid structural system used to support cables. Cable tray systems include ladders, troughs, channels, solid bottom trays, and other similar structures.

Cablebus. An approved assembly of insulated conductors with fittings and conductor terminations in a completely enclosed, ventilated, protective metal housing.

Capable of being locked out. The capability of an energy isolation device to have a lock attached to secure the energy source in a "safe" position (OFF, OPEN, or CLOSED). The "safe" position is the position that will isolate the energy source, relieve pressure, or prevent an unsafe accumulation of energy.

Circuit breaker. (600 volts nominal, or less). A device designed to open and close a circuit by non-automatic means and to open the circuit automatically on a predetermined overcurrent without injury to itself when properly applied within its rating.

(Over 600 volts, nominal). A switching device capable of making, carrying, and breaking currents under normal circuit conditions, and also making, carrying for a specified time, and breaking currents under specified abnormal circuit conditions, such as those of short circuit.

Class I locations. Locations in which flammable gases or vapors are or may be present in the air in quantities sufficient to produce explosive or ignitable mixtures. Class I locations include the following:

Class I, Division 1. Classification for a location: (a) in which hazardous concentrations of flammable gases or vapors may exist under normal operating conditions; or (b) in which hazardous concentrations of such gases or vapors may exist frequently because of repair, maintenance operations, or leakage; or (c) in which breakdown or faulty operation of equipment or processes might release hazardous concentrations of flammable gases or vapors, and might also cause simultaneous failure of electric equipment.

Class I, Division 2. Classification for a location: (a) in which volatile flammable liquids, vapors, or gases are handled, processed, or used—but will normally be confined within closed containers or closed systems from which they can escape only in case of accidental rupture or breakdown of such containers or systems, or in case of abnormal operation of equipment; or (b) in which hazardous concentrations of gases or vapors are normally prevented by positive mechanical ventilation, and which might become hazardous through failure or abnormal operations of the ventilating equipment; or (c) that is adjacent to a Class I, Division 1 location which might occasionally communicate hazardous concentrations of gases or vapors, unless such communication is prevented by adequate positive-pressure ventilation from a source of clean air, and effective safeguards against ventilation failure are provided.

Class II locations. Locations that are hazardous because of the presence of combustible dust. Class II locations include the following:

Class II, Division 1. Classification for a location: (a) in which combustible dust is or may be in suspension in the air under normal operating conditions, in quantities sufficient to produce explosive or ignitable mixtures; or (b) where mechanical failure or abnormal operation of machinery or equipment might cause such explosive or ignitable mixtures to be produced and might also provide a source of ignition through simultaneous failure of electric equipment, operation of protection devices, or from other causes, or (c) in which combustible dusts of an electrically conductive nature may be present.

Class II, Division 2. Classification for a location in which: (a) combustible dust will not normally be in suspension in the air in quantities sufficient to produce explosive or ignitable mixtures, and dust accumulations are normally insufficient to interfere with the normal operation of electrical equipment or other apparatus; or (b) dust may be in suspension in the air as a result of infrequent malfunctioning of handling or processing equipment, and the resulting dust accumulations may be ignitable by abnormal operation or failure of electrical equipment or other apparatus.

Class III locations. Locations that are hazardous due to the presence of easily ignitable fibers or flyings but in which such fibers or flyings are not likely to be in suspension in the air in quantities sufficient to produce ignitable mixtures. Class III locations include the following:

Class III, Division 1. Classification for a location in which easily ignitable fibers or materials producing combustible flyings are handled, manufactured, or used.

Class III, Division 2. Classification for a location in which easily ignitable fibers are stored or handled, except in process of manufacture.

Damp location. Partially protected locations under canopies, marquees, roofed open porches, and like locations; interior locations subject to moderate degrees of moisture, such as some basements, some barns, and some cold-storage warehouses.

Disconnecting means. A device, group of devices, or other means by which the conductors of a circuit can be disconnected from their source of supply.

Disconnecting (or Isolating) switch. (Over 600 volts, nominal.) A mechanical switching device used for isolating a circuit or equipment from a source of power.

Dry location. A location not normally subject to dampness or wetness. A location classified as dry may be temporarily subject to dampness or wetness, as in the case of a building under construction.

Energized. A condition in which a machine is connected to an energy source (electrical, pneumatic, hydraulic) or contains stored energy (electric charge, magnetic field, elevated position, compressed or stretched spring, pressurized liquid or gas).

Energy isolation device. A device that provides positive, complete, reliable, and verifiable isolation from an energy source. Such devices are mechanical valves with position indicators and electric circuit breakers and fuses. Valves without position indicators or electrically driven valves are not acceptable energy isolation devices, unless the position of the valve can be verified and all energy to the valve circuitry can be isolated. Electrical control switches are not acceptable energy isolation devices because they have many failure modes and frequent use makes them susceptible to fatigue failure. Furthermore, many electrical control switches do not function exactly as labeled. Many circuits remain energized even when control switches are in the OFF position.

Equipment Grounding Conductor. The conductor used to connect the non-current-carrying, metal parts of equipment, raceways, and other enclosures to the system's grounded conductor and/or the grounding electrode conductor at the service equipment or at the source of a separately derived system.

Exposed. (As applied to live parts.) Capable of being inadvertently touched or approached nearer than a safe distance by a person. It is applied to parts not suitably guarded, isolated, or insulated.

Exposed. (As applied to wiring methods.) On or attached to the surface or behind panels designed to allow access.

Fixed wiring. Wiring fastened and supported in place so that it cannot move.

Feeder. All circuit conductors between the service equipment or the generator switchboard of an isolated plant, and the final branch-circuit overcurrent device.

Flexible wiring. Fixed wiring made of flexible conduit or tubing that has been fastened and supported in place so that it cannot move.

Fuse. (Over 600 volts, nominal.) An overcurrent protective device with a circuit opening fusible part that is heated and severed by the passage of overcurrent through it. A fuse comprises all the parts that form a unit capable of performing the prescribed functions. It may or may not be the complete device necessary to connect it into an electrical circuit.

Flyings. Light-weight, fibrous material or agglomerates of fibers that are easily dispersed and suspended in air.

Ground. A conducting connection, whether intentional or accidental, between an electrical circuit or equipment and the earth, or to some conducting body that serves in place of the earth.

Grounded. Connected to earth or to some conducting body that serves in place of the earth.

Grounded conductor. A system or circuit conductor that is intentionally grounded.

Grounding conductor. A conductor used to connect equipment or the grounded circuit of a wiring system to a grounding electrode or electrodes.

Ground Fault Circuit Interrupter. A device whose function is to interrupt the electric circuit to the load when a fault current to ground exceeds some predetermined value that is less than that required to operate the overcurrent protective device of the supply circuit.

Hoistway. Any shaft, hatchway, well hole, or other vertical opening or space in which an elevator or dumbwaiter is designed to operate.

Interchange. Removing, installing, or changing out one appliance or fixture for another. Flexible cables are used to make frequent interchange of appliances or fixtures easy and fast.

Labeled. Designation for equipment attached with a label, symbol, or other identifying mark of a nationally recognized testing laboratory which: (a) makes periodic inspections of the production of such equipment, and (b) whose labeling indicates compliance with nationally recognized standards or tests to determine safe use in a specified manner.

Listed. Designation for equipment mentioned in a list which: (a) is published by a nationally recognized laboratory which makes periodic inspection of the production of such equipment, and (b) states such equipment meets nationally recognized standards or has been tested and found safe for use in a specified manner.

Lockout device. A device that utilizes a positive means, such as a lock (either key or combination type) to hold an energy isolation device in the safe position and prevent the energizing of a machine or equipment. Lockout devices include blank flanges and bolted slip blinds.

Lockout/tagout. The control of hazardous energy during maintenance or servicing in which all energy sources are isolated and locked or tagged, all stored energy is released and prevented from accumulating, and a zero energy state is verified.

Maintenance or servicing. Work on equipment or machinery that includes installation, setup, adjustment, inspection, lubrication, cleaning, clearing of jammed machinery, and tool changes. If the servicing is minor, routine, repetitive, and integral to the use of the equipment for production, the lockout/tagout standard and procedures do not apply provided that the work is performed using alternative measures that provide effective protection.

MCC. See Motor Control Cabinet.

Metric Units. Units of physical parameters agreed to by international committee and adopted by individual countries. The units are based on factors of 10 with the following prefixes:

Giga- (G)	10^9
Mega- (M)	10^6
Kilo- (k)	10^3
Centi- (c)	10^{-2}
Milli- (m)	10^{-3}

Motor Control Cabinet (MCC). The electrical enclosure containing the motor control circuitry including the main power relay for the motor.

National Fire Protection Association (NFPA). A private organization that publishes industry consensus standards, such as the National Electrical Code, that are adopted by federal, state, and local regulations.

National Electrical Code (NEC). A standard published by the National Fire Protection Association for the safe design and installation of electrical premises wiring. Some edition of this standard is usually adopted as the local or state electrical code throughout the United States.

National Electrical Safety Code (NESC). A standard published by the American National Standards Institute for the safe design, installation, and operation or power generation, transmission, and distribution equipment.

NEC. See National Electrical Code.

NESC. See National Electrical Safety Code.

NFPA. See National Fire Protection Association.

Occupational Safety and Health Administration (OSHA). The federal agency responsible for employee safety and health. Many states have their own employee safety and health regulations which must be at least as protective as federal regulations. Some states have federally-approved state enforcement agencies that enforce the state regulations instead of federal regulators.

Outlet. An outlet where one or more receptacles are installed.

Overcurrent. Any current in excess of the rated current of equipment or the listed current-carrying capability of a conductor. It may result from overload (see definition), short circuit, or ground fault.

OSHA. See Occupational Safety and Health Administration.

Overload. Operation of equipment in excess of normal, full load rating, or of a conductor in excess of its rated current-carrying capability which, if operated in this manner for a sufficient length of time, would cause damage or dangerous overheating. A fault, such as a short circuit or ground fault, is not an overload. (See Overcurrent.)

Panelboard. A single panel or group of panel units designed for assembly in the form of a single panel. A panelboard includes buses and automatic overcurrent devices and is equipped with or without switches for the control of light, heat, and power circuits. It is designed to be placed in a cabinet or cutout box placed in or against a wall or partition and accessible only from the front.

PLC. See Programmable Logic Controller.

Premises wiring. Interior and exterior wiring, including power, lighting, control, and signal circuit wiring—together with all associated hardware, fittings, and wiring devices, both permanently and temporarily installed. Premises wiring extends from the load end of the service drop or load end of the service lateral conductors to the outlet(s). Such wiring does not include wiring internal to appliances, fixtures, motors, controllers, motor control centers, or similar equipment.

Programmable Logic Controller (PLC). A device that stores software instructions for the computer control of equipment and machines.

Qualified employee. One familiar with the construction and operation of the equipment and the hazards involved.

Raceway. A channel designed expressly for holding wires, cables, or busbars, with additional functions as permitted in this subpart. Raceways may be of metal or insulating material, and the term includes rigid metal conduit, rigid nonmetallic conduit, intermediate metal conduit, liquidtight flexible metal conduit, flexible metallic tubing, flexible metal conduit, electrical metallic tubing, underfloor raceways, cellular concrete floor raceways, cellular metal floor raceways, surface raceways, wireways, and busways.

Radiofrequency (RF). Electromagnetic, non-ionizing radiation in the range of 30 kilo-Hertz to 300 giga-Hertz.

Readily accessible. Capable of being reached quickly for operation, renewal, or inspection without requiring one to climb over obstacles, remove obstacles, or resort to the help of portable ladders or chairs.

Receptacle. A receptacle is a contact device installed at the outlet for the connection of a single attachment plug. A single receptacle is a single contact device with no other contact device on the same yoke. A multiple receptacle is a single device containing two or more receptacles.

Rigid wiring. Fixed wiring made of rigid or stiff conduit that holds its shape. Some conduit can be bent into a desired shape. Some conduit has to be assembled with pre-formed elbows to achieve desired shape.

Service. The conductors and equipment for delivering energy from the electricity supply system to the wiring system of the premises served.

Service drop. The overhead service conductors from the last pole or other aerial support to and including the splices, if any, connecting to the service-entrance conductors at the building or other structure.

Switchboard. A large single panel, frame, or assembly of panels on which are mounted, on the face or back, or both, switches, overcurrent and other protective devices, buses, and usually instruments. Switchboards are generally accessible from the rear as well as from the front and are not intended to be installed in cabinets.

Switchgear. A large single panel, frame, and enclosure containing switches, interrupting devices and their control, metering, protection, and regulating equipment where integral to the assembly, with associated interconnections and supporting structures. Such switchgear is used in substations, power centers, service points, or similar equipment.

Tagout device. A prominent warning device, such as a tag and its means of attachment, which can be securely fastened to an energy isolation device in accordance with an established procedure, to indicate that the energy isolation device and the equipment being controlled may not be operated until the tagout device is removed.

Threshold Limit Value (TLV). An 8-hour, time-weighted, average exposure limit published by the American Conference of Governmental Industrial Hygienists.

TLV. See Threshold Limit Value.

Wet location. Installations underground or in concrete slabs or masonry in direct contact with the earth, and locations subject to saturation with water or other liquids, such as vehicle-washing areas, and locations exposed to weather and unprotected.

Wireways. Sheet-metal troughs with hinged or removable covers for housing and protecting electric wires and cable and in which conductors are laid in place after the wireway has been installed as a complete system.

Government Institutes Mini-Catalog

PC #	ENVIRONMENTAL TITLES	Pub Date	Price*
629	ABCs of Environmental Regulation	1998	$65
672	Book of Lists for Regulated Hazardous Substances, 9th Edition	1999	$95
4100	CFR Chemical Lists on CD ROM, 1999-2000 Edition	1999	$125
512	Clean Water Handbook, Second Edition	1996	$115
581	EH&S Auditing Made Easy	1997	$95
673	E H & S CFR Training Requirements, Fourth Edition	2000	$99
825	Environmental, Health and Safety Audits, 8th Edition	2001	$115
548	Environmental Engineering and Science	1997	$95
643	Environmental Guide to the Internet, Fourth Edition	1998	$75
820	Environmental Law Handbook, Sixteenth Edition	2001	$99
688	EH&S Dictionary: Official Regulatory Terms, Seventh Edition	2000	$95
821	Environmental Statutes, 2001 Edition	2001	$115
4099	Environmental Statutes on CD ROM for Windows-Single User, 1999 Ed.	1999	$169
707	Federal Facility Environmental Compliance and Enforcement Guide	2000	$115
708	Federal Facility Environmental Management Systems	2000	$99
689	Fundamentals of Site Remediation	2000	$85
515	Industrial Environmental Management: A Practical Approach	1996	$95
510	ISO 14000: Understanding Environmental Standards	1996	$85
551	ISO 14001: An Executive Report	1996	$75
588	International Environmental Auditing	1998	$179
518	Lead Regulation Handbook	1996	$95
608	NEPA Effectiveness: Mastering the Process	1998	$95
582	Recycling & Waste Mgmt Guide to the Internet	1997	$65
615	Risk Management Planning Handbook	1998	$105
603	Superfund Manual, 6th Edition	1997	$129
685	State Environmental Agencies on the Internet	1999	$75
566	TSCA Handbook, Third Edition	1997	$115
534	Wetland Mitigation: Mitigation Banking and Other Strategies	1997	$95

PC #	SAFETY and HEALTH TITLES	Pub Date	Price*
697	Applied Statistics in Occupational Safety and Health	2000	$105
547	Construction Safety Handbook	1996	$95
553	Cumulative Trauma Disorders	1997	$75
663	Forklift Safety, Second Edition	1999	$85
709	Fundamentals of Occupational Safety & Health, Second Edition	2001	$69
612	HAZWOPER Incident Command	1998	$75
662	Machine Guarding Handbook	1999	$75
535	Making Sense of OSHA Compliance	1997	$75
718	OSHA's New Ergonomic Standard	2001	$95
558	PPE Made Easy	1998	$95
683	Product Safety Handbook	2001	$95
598	Project Mgmt for E H & S Professionals	1997	$85
658	Root Cause Analysis	1999	$105
552	Safety & Health in Agriculture, Forestry and Fisheries	1997	$155
669	Safety & Health on the Internet, Third Edition	1999	$75
668	Safety Made Easy, Second Edition	1999	$75
590	Your Company Safety and Health Manual	1997	$95

Government Institutes
4 Research Place, Suite 200 • Rockville, MD 20850-3226
Tel. (301) 921-2323 • FAX (301) 921-0264
Email: giinfo@govinst.com • Internet: http://www.govinst.com

Please call our customer service department at (301) 921-2323 for a free publications catalog.

CFRs now available online. Call (301) 921-2355 for info.

*All prices are subject to change. Please call for current prices and availablity.

Government Institutes Order Form

4 Research Place, Suite 200 • Rockville, MD 20850-3226
Tel (301) 921-2323 • Fax (301) 921-0264
Internet: http://www.govinst.com • E-mail: giinfo@govinst.com

4 EASY WAYS TO ORDER

I. Tel: **(301) 921-2323**
Have your credit card ready when you call.

2. Fax: **(301) 921-0264**
Fax this completed order form with your company
purchase order or credit card information.

3. Mail: **Government Institutes Division**
ABS Group Inc.
P.O. Box 846304
Dallas, TX 75284-6304   USA

Mail this completed order form with a check, company purchase
order, or credit card information.

4. Online: Visit http://www.govinst.com

PAYMENT OPTIONS

❏ **Check** *(payable in US dollars to **ABS Group Inc. Government Institutes Division**)*

❏ **Purchase Order** *(This order form must be attached to your company P.O.* <u>Note:</u>*All International orders must be prepaid.)*

❏ **Credit Card**   ❏ VISA   ❏ MasterCard   ❏ AMERICAN EXPRESS

Exp. ___ /____

Credit Card No. _____

Signature _____

(Government Institutes' Federal I.D.# is 13-2695912)

CUSTOMER INFORMATION

Ship To: (Please attach your purchase order)

Name _____

GI Account # *(7 digits on mailing label)* _____

Company/Institution _____

Address _____
(Please supply street address for UPS shipping)
_____

City _____ State/Province _____

Zip/Postal Code _____ Country _____

Tel (____) _____

Fax (____) _____

E-mail Address _____

Bill To: (if different from ship-to address)

Name _____

Title/Position _____

Company/Institution _____

Address _____
(Please supply street address for UPS shipping)

City _____ State/Province _____

Zip/Postal Code _____ Country _____

Tel (____) _____

Fax (____) _____

E-mail Address _____

Qty.	Product Code	Title	Price

30 DAY MONEY-BACK GUARANTEE

If you're not completely satisfied with any product, return it undamaged
within 30 days for a full and immediate refund on the price of the product.

Subtotal _____
MD Residents add 5% Sales Tax _____
Shipping and Handling (see box below) _____
Total Payment Enclosed _____

SOURCE CODE: BP03

Shipping and Handling	**Sales Tax**
Within U.S:	Maryland 5%
1-4 products: $6/product	Texas 8.25%
5 or more: $4/product	Virginia 4.5%
Outside U.S:	
Add $15 for each item (Global)	